CARVING GRAND CANYON

EVIDENCE, THEORIES, and MYSTERY

WAYNE RANNEY

GRAND CANYON ASSOCIATION

Edited by Pam Frazier
Designed by Larry Lindahl
Maps and diagrams by Bronze Black
Indexed by Wayne Ranney
Printed in Canada by
Friesens & Sons on recycled paper
using vegetable-based inks.

Reprint History 10 9 8 7 6 5 4 3 2

ISBN 0-938216-82-1
Library of Congress Control Number:
2005000477

Grand Canyon Association is a non-profit educational organization working in partnership with Grand Canyon National Park since 1932. Net proceeds from the sale of this book will be used to support the education and research goals of Grand Canyon National Park.

Grand Canyon Association
P.O. Box 399
Grand Canyon, AZ 86023-0399
(800) 858-2808
www.grandcanyon.org

Library of Congress Cataloging-in-Publication Data

Ranney, Wayne.
Carving Grand Canyon / by Wayne Ranney.— 1st ed.
p. cm.
Includes bibliographical references and index.
ISBN 0-938216-82-1 (pbk.)
1. Geology—Arizona—Grand Canyon—Popular works. 2. Geology—Colorado River (Colo.-Mexico)—Popular works. 3. Grand Canyon (Ariz.)—Discovery and exploration—Popular works. I. Title.
QE86.G73R36 2005
557.91'32—dc22

2005000477

Wayne Ranney

Contents

Cover: Sunset on the Colorado River near Nankoweap
Photograph by Larry Lindahl

Title Page: View west from Yaki Point
Photograph by George H. H. Huey

Table of Contents: Detail from *Point Sublime* by William H. Holmes, 1882

ACKNOWLEDGMENTS

This book benefited greatly from the help and encouragement of many people. First and foremost, I thank all of my students at Yavapai College in both Prescott and Sedona who have encouraged me through the years to put my classroom lectures into written form. I will always be grateful for your support.

I thank the Grand Canyon Association for their trust that I could tell this story. Pam Frazier especially, guided the editorial process with professionalism, friendship, and above all, grace. I thank the Museum of Northern Arizona and the Grand Canyon Field Institute for continuing to provide me with opportunities to take backpackers and river runners into the canyon.

Stewart Aitchison gave me my first opportunity to become an international guide and continually encouraged me to put my Grand Canyon origin talks into book form. Thank you Stewart. Amy Rosen, Cassie Fenton, and Jill Robertson were among those who reviewed portions of the text. And many thanks to the Frampton, Mistretta, and Nelson families whose questions about the Grand Canyon, while hiking the Highline Trail in the Canadian Rockies one fine August day in 2003, inspired the book's summary chapter.

My thanks also to my fellow Grand Canyon geologists and guides whose hard work and dedication have helped me to know the Grand Canyon better. Among this group I give special thanks to Ivo Lucchitta whose thoughtful insights and excellent speaking and writing skills have been an inspiration to me since I first came to know the Grand Canyon in 1975.

Lastly, I would like to acknowledge the first geologists to write Grand Canyon origin books for a general audience: N. H. Darton, 1917; Eddie McKee, 1931; and John Maxson, 1962. I hope this work carries on in the spirit and tradition they began.

DEDICATION

This book is dedicated to the Museum of Northern Arizona and the Grand Canyon Association, who together published the "Geologic Map of the Eastern Part of the Grand Canyon National Park, Arizona" in 1976. This colorful map, which has been a fixture on the walls of my home since its publication, has served to remind me every day that the Grand Canyon and all that it represents is there always to inspire and enchant. A unique blend of science and art, this map has been the primary motivator for my interest in Grand Canyon's origin and without its publication this book might never have seen the light of day. Both the Museum of Northern Arizona and the Grand Canyon Association, whose missions are to increase and share knowledge about the Grand Canyon region, deserve every ounce of support we can give them.

Preface

I love the Grand Canyon! For me there is no more powerful place on earth than this huge chasm of stone and light. Since 1975, when I arrived here as a wide-eyed, twenty-one-year-old boy, the canyon and the surrounding landscape have remained the recreational, intellectual, and spiritual focus of my life. I am fortunate that my work as a geologist, teacher, and guide has allowed me to repeatedly enter the canyon, where I have developed an intimate relationship with the sweeping ramparts and vast silence of this great excavated space. I am grateful to know the Grand Canyon so well; its lessons have engendered in me the deepest respect for it and the ideas related to its mysterious evolution. It has served as the perfect classroom to teach geology but also it has been a teacher and mentor to me. I have become a geologist because of its lessons in earth history but more important, I have come of age as a human being in the depths of its salubrious embrace. In a very real way I am a child of the Grand Canyon.

In June 2000, I was privileged to participate in the Symposium on the Origin of the Colorado River held at Grand Canyon Village. Over seventy geologists attended and thirty-six scientific abstracts were presented in a working volume that spoke to the origin of the river and the canyon. The primary sponsor of the conference, the Grand

Above:
Bright Angel Point
oil on canvas
by Bruce A. Aiken, 1995

Opposite:
Detail from *Geologic Map of the Eastern Part of Grand Canyon National Park* Published by the Grand Canyon Association, 1996

Canyon Association, published a monograph in 2004 that was written and compiled for use by professional geologists. Although eminently up-to-date regarding most of the current theories, the monograph was not intended to be used or even necessarily understood by a broader audience, who oftentimes may wonder about the canyon's formation while looking into its colorful and wondrous depths.

This prompted me to pursue an idea that has continued to preoccupy me as I take visitors down into the Grand Canyon each year. Since 1975 I have led people of various backgrounds on multi-day backpacks, river trips, or rim tours, providing lectures on the possible origin of the canyon. These lectures, given spontaneously at first, were difficult to organize since I was trying to convey the many nuances that are inherent in such a complex topic. It was unsettling to see the confusion on some faces as I realized halfway through a talk that I had forgotten to mention a basic fact of the river's evolution or the canyon's history.

Compelled by the audience and the Grand Canyon itself, I continued to refine my thoughts and my way of sharing them so that they could be increasingly understood. The symposium in 2000 provided me with an incentive to write this book. It also gave me a framework of ideas that I have strived to synthesize into a more understandable story for readers such as yourself, who may be interested in the story of the canyon but may not be familiar with the jargon of geology. I have increasingly received heartfelt encouragement from many friends, colleagues, students, and visitors to put my lectures in written form so that a larger audience could share in the fantastic story of the formation of the Colorado River and its Grand Canyon.

Humility, respect, curiosity, unbounded enthusiasm, and ever increasing circles of earthly and personal exploration are just some of the life lessons that this living gorge has taught me. The Grand Canyon has shaped and guided my life. It is a distinct honor for me to share its story with you and I hope you will find it as compelling and interesting as I do.

Late afternoon in the Grand Canyon
Photograph by Tom Till

Introduction

THE GRAND CANYON is a world unto itself! When one turns his or her back to the plateau that ends abruptly at its edge and descends a Grand Canyon trail, they enter an ethereal world of light and rock. It is a landscape very much coherent and integrated within itself, yet separate and distinct from the one that surrounds it. To some it may appear as if in ruins, perhaps even unfinished, as the jagged cliffs yield ever more boulders to the slopes below. Yet it is probably true that its depth or profile has not changed significantly since the end of the last Ice Age some ten thousand years ago. In many ways, the Grand Canyon is an enigma but one that ultimately inspires and creates the greatest awe within the souls of those who view it.

This canyon is one of our planet's most sublime and spectacular landscapes, yet to this day it defies complete understanding of how it came to be. It is visited by millions of people a year and not one of them knows precisely how or when it formed. The canyon's birth is shrouded in hazy mystery, cloaked in intrigue, and filled with enigmatic puzzles. And although the Grand Canyon is one of the world's most recognizable landscapes, it is remarkable how little is known about the details of its origin.

This may come as a surprise to those who know that geologists speak with confidence about certain aspects of

Above:
The Inner Gorge
Acrylic on board
by Bruce A. Aiken, 1988

Opposite:
The Grand Canyon is a source of inspiration and enchantment for many who fall under its spell.
Photograph by Gary Ladd

THE GRAND CANYON OF THE COLORADO RIVER

Portrayed in this digital Landsat image, the Grand Canyon is perhaps our planet's greatest example of the erosive force of running water. More than three hundred miles of river are visible here from Lake Powell (upper right) to the Grand Wash Cliffs at Lake Mead (center left). On page 13, note the high crest of the Kaibab Plateau made visible by snow cover north of the gorge. Landsat image courtesy of U.S. Geological Survey, Southwest Geographic Science Team, Flagstaff Office.

Vermilion Cliffs
Echo Cliffs
Marble Canyon
KAIBAB
PLATEAU
N
Little Colorado
River
COCONINO
PLATEAU
San Francisco Peaks

our planet's ancient history and the life forms that once lived here. We routinely search the far reaches of our solar system for planetary volcanoes, Martian water, or frozen seas of methane. Yet until recently our own planet's most famous and most visited "hole in the ground" rarely received a word about how it may have formed. How can one of the most treasured landscapes on Earth escape a unifying theory regarding its formation?

The foremost reason why this is true is that the Grand Canyon has been shaped largely by erosion. As the Colorado River has continued to deepen its track, other erosive forces widen the canyon removing most of the evidence for its early incarnation. As the canyon has become deeper and wider, its history has receded into the shadowy depths, much like the cliffs that enclose it when the sun sinks below the rim. We may never know the intimate details of Grand Canyon's origin because most of the evidence has been washed downstream and is gone forever.

Another possible reason why there is no single coherent theory about Grand Canyon's origins is because there has been no economic incentive to understand it. Volumes have been written about obscure geologic features found in adjacent areas where the economic rewards are great—gold, uranium, coal, oil, and gas.

Rivers such as the Colorado and the Little Colorado (pictured here) carry large volumes of eroded sediment, removing evidence of earlier geologic development of the canyon. Photograph by Mike Buchheit

There is no financial reward at the end of the Grand Canyon story and, partly for this reason, many geologists have left it alone. Amazingly, if a relatively small amount of money used in unraveling the intricacies of a nearby oil field were redirected into a scientific study of the evolution of the Colorado River, we might have a better understanding of its origin and development in a much shorter time.

However, the one aspect of its character that may be most responsible for the lack of a unifying theory is its immense size with respect to both time and space. The Grand Canyon is simply too big to be known easily. It has existed through so vast a time that our short, human frame of reference cannot possibly grasp so monumental a puzzle. Its history straddles many distinct geologic events that have created much of our western landscape, adding layer upon layer of complexity and doubt. Imagine, it is so vast that various parts of the canyon may have formed at completely different times!

Regarding its size, consider the following: measured along the river the canyon is 277 miles long. Arguably, there is no other single canyon in the world that approaches its length. A vehicle traveling along an imaginary highway parallel to the Colorado River at the eye blurring speed of seventy miles an hour would take nearly four hours to drive through the canyon! The greater portion of this length is a mile deep within the earth, with individual sheer escarpments that are three hundred, five hundred, even three thousand feet in height. It requires phenomenal endurance and stamina to explore this rugged defile which only a few hardy souls can muster during their lifetime. It is Grand Canyon's immense size and rough terrain that has kept most geologists confined to one or both of her two rims, her few trails, or single river corridor.

The amount of time that the Colorado River has been carving Grand Canyon is the subject of much controversy and debate. Geologists who have studied the canyon during their entire careers can only suggest a range of ages somewhere between 70 and 6 million years, although the younger age is most widely accepted. This range of dates is perhaps testament to how unknowable the canyon really is to us. And while the rocks enclosing Grand Canyon stretch back as much as twenty-five times longer than the oldest events even peripherally related to the river's history, 70 million years is still an incredibly long time to comprehend.

In addition to the reasons cited above, we find ourselves at a loss to define exactly what the Grand Canyon is through time. For example, if the canyon were only 500 feet deep and only one mile in width, as it must have been at some point early in its history, would that still be the Grand Canyon? What if it were born in this exact place but within higher rock strata that are now completely eroded from this area? What if it were born from a river that went in the opposite direction and thus was totally unrecognizable to us as the Colorado? Would that still be the Grand Canyon? As we attempt to understand how and when the Grand Canyon may have formed, we must remember that defining when it began could become a game defining what it is.

The mysteries that surround the origin of the Colorado River in Grand Canyon contribute tremendously to its fascination and charm. Those who have spent decades inside the gorge contemplating the possible configurations of the river's ancient course may one day feel a significant loss should someone develop a theory accepted as truth concerning its origin. Yet many try—not only geologists but river runners, hikers, park rangers, cartographers, and dreamers. I am glad to have lived a good portion of my life in a time when the canyon is not fully understood. It is comforting to know that there will always be significant parts of its history that will remain forever unreachable and unknowable to us.

The story you are about to read has no definitive answer at the end, no "a-ha!" moment. We are frequently left with more questions than answers simply because the river continues to excavate away at the traces of its early history, leaving us behind in bewilderment as we scratch our heads in disbelief. There are few places where one can go to learn how the Grand Canyon formed. There are no interpretive signs inside the park that speak to the idea of how the canyon may have formed and most books say little more than the river carved the canyon. This is the first book in recent times written for a general audience that focuses specifically on how the Grand Canyon may have formed. And although it may not be the last word on the topic, it is my hope that this book will serve as a way for interested visitors to know more about the details of Grand Canyon's possible origins. As an educator, I understand that there are many ways to tell a story and wholeheartedly welcome attempts by others to tell it their way. I ask your indulgence for the way I have told it.

I present the known story of the canyon's formation in three ways: first, by presenting ideas on how rivers in general can physically carve canyons; second, by looking chronologically at the numerous theories that have been presented by successive generations of geologists; and third, by describing a plausible sequence of geologic events that could create this landscape. The first part tells of the processes that have been at work to create the canyon; the second part takes us on a collective journey through the last one hundred fifty years of scientific inquiry; and the third, suggests how and when it may have formed. The reader may follow these ideas as they were presented, discussed, revised, or discarded by those who spent considerable time and effort studying the canyon's origins. The book is intended to be not only a trip through geologic time but one through human history as well, as humans have attempted to unravel the immense mystery and philosophical depth of this wondrous gorge.

Let us take a journey then—a journey through time along the Colorado River in Grand Canyon. We will witness ancient rivers that were born in the wake of retreating seas, experience crustal uplift as our continent was squeezed and compressed, and discuss ways in which a river may reverse its course or how a canyon is deepened. Let's journey with John Wesley Powell, Clarence Dutton, and a host of modern scientists who remain forever fascinated by the Grand Canyon of the Colorado River.

Detail from *Point Sublime* by William H. Holmes, 1882

1

The Enigma of Grand Canyon

Grand Canyon is a puzzle, a mystery, an enigma. It appears to have been carved through an uplifted plateau, ignores fault lines, may have been born by a river that once flowed the other way, is possibly quite old or quite young—or both—and is set within a more mature landscape.

The Run at Horn Creek Rapids
Oil on canvas
by Bruce A. Aiken, 1998

SOME PEOPLE may wonder why there is still so much controversy among geologists concerning ideas on the origin of the Grand Canyon. Perhaps they suspect that in our modern world, with all of its technologic and scientific advances, questions about the canyon's history have been fully answered. They oftentimes seem surprised to learn that geologists still refer to ongoing "problems" associated with understanding the details of Grand Canyon's origin. "Didn't the river carve it?" people invariably ask. The answer is absolutely yes and the one truth that every geologist agrees upon is that the Colorado River carved the Grand Canyon. But more important are the deeper questions: "How did the river cut the canyon?" "When did it accomplish its task and by which manner of erosion?" Geologists remain perplexed by these more difficult questions and continue to

Grand Teton

Yellowstone

Yosemite

The geologic evolution of national parks such as Grand Teton, Yellowstone, and Yosemite are better understood, and the theories are less controversial, than Grand Canyon's. Photographs by Wayne Ranney

puzzle over the subtle intricacies and lack of meaningful clues about how and when this landscape evolved.

Grand Canyon is somewhat unique among our national parks because of the lack of a single, scientific theory regarding its origin. The Grand Tetons are known to be one of the youngest mountain ranges in all of the Rocky Mountain system, having been uplifted above the valley of Jackson Hole within the last 13 million years. When visitors inquire about the Yellowstone landscape, they are told about the catastrophic volcanic eruptions during the last 2 million years that have left the world's largest concentration of geysers, hot springs, fumaroles and mud pots in their wake. And Yosemite Valley displays clear evidence of being reshaped by glaciers during the last Ice Age. But, Grand Canyon's origin remains shrouded in mystery and there are few places visitors can go to obtain even a rudimentary understanding about its beginnings. This lack of public information speaks more about the canyon's enigmatic setting within the plateau landscape than some larger conspiracy to keep people in the dark about such matters.

Oddly enough, the Grand Canyon is located in a place where it seemingly shouldn't be. Some twenty miles east of Grand Canyon Village the Colorado River turns sharply ninety degrees, from a southern course to a western one and into the heart of the uplifted Kaibab Plateau. Under ordinary circumstances an uplifted plateau acts as a barrier to a river's course, causing it to flow around that barrier through lower ground. Rivers do not normally flow into elevated plateaus but the Colorado River is not a normal river. It appears to cut right through this uplifted wall of rock, which lies three thousand feet above the adjacent Marble Platform to the east. This odd scenario was the foremost problem recognized by the very

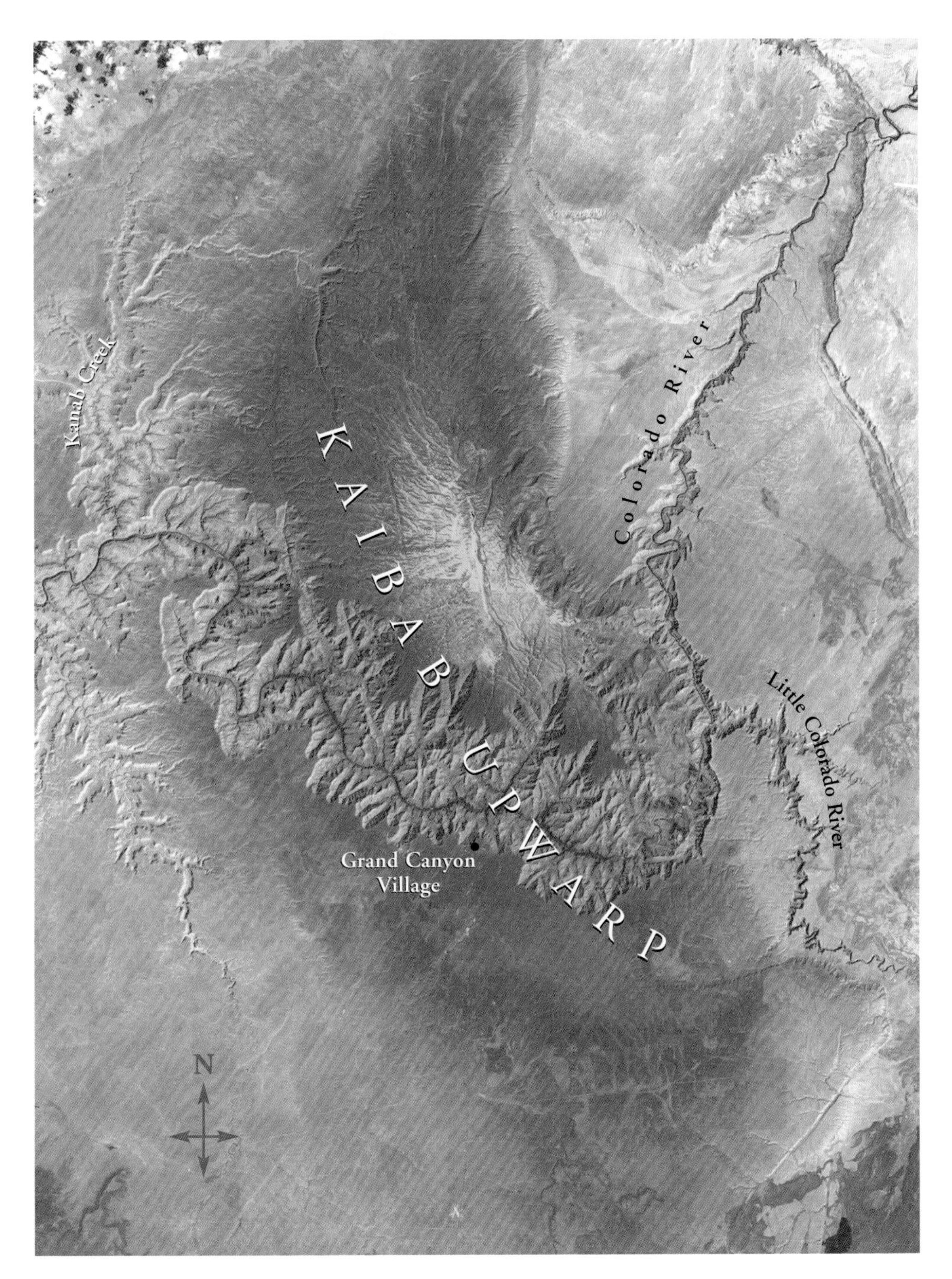

The north-to-south course of the Colorado River turns abruptly west cutting into the high-elevation, forested Kaibab upwarp (dark green). This odd placement is among the foremost puzzles in understanding the origin of the Colorado River and the Grand Canyon.

Landsat image courtesy of U.S. Geological Survey, Southwest Geographic Science Team, Flagstaff Office

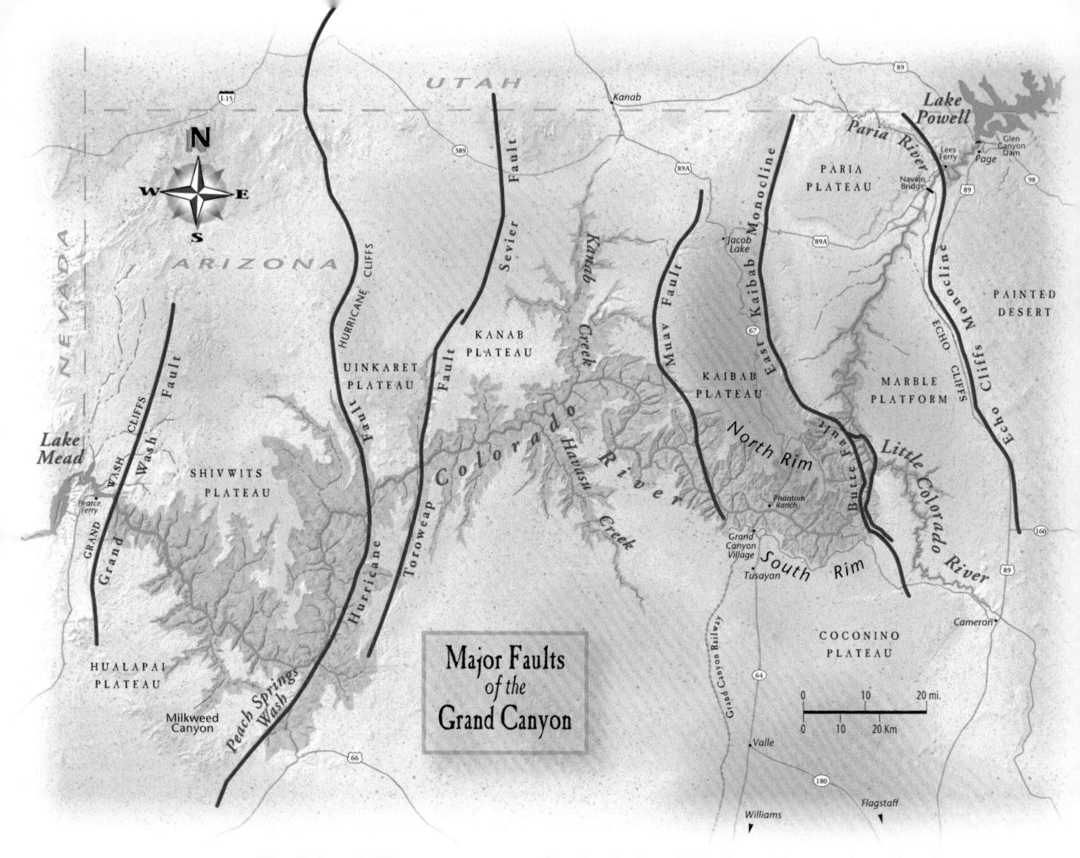

The Colorado River crosses several major faults within Grand Canyon, ignoring these zones of weakness in the landscape and prompting questions about the river's origin.

first geologists who saw the Grand Canyon. Why does the Colorado River seem to flow into the heart of an uplifted plateau?

Another curiosity with the Colorado River's course is that it disregards the fault lines that cross its path. Rivers oftentimes follow faults where repeated earthquakes have broken and pulverized the ground, creating linear zones of weakened rock. Erosion by a river is more easily accomplished along these lines of shattered rock. The Colorado River within Grand Canyon crosses dozens of faults, many of them at right angles, and continues on downstream through blocks of strata that are solid and unbroken by faults. Although there is one twenty-five-mile stretch where the river parallels the Hurricane Fault, this is the exception rather than the rule. Why does the Colorado River in Grand Canyon disregard the faults that cross its path, lines that seemingly offer less resistance?

Another fact that begs for explanation is that in some sections of the canyon the patterns and direction of drainage suggest that the

Colorado River could have once flowed in the opposite direction. Ancient gravel layers found near the Grand Canyon reveal evidence for this. Additionally, in Marble Canyon, many tributary streams come into the Colorado River flowing generally to the north, against the southerly flow of the modern river. This creates a pattern of drainage known to geologists as "barbed" tributaries. The Marble Platform, into which the tributaries have been carved, also slopes down to the northeast exactly opposite the flow direction of the modern river. The tributaries point in an arrow-like manner to a postulated previous flow direction of the Colorado. Could the Colorado River have once flowed north through this or other parts of Grand Canyon, only later to have reversed its flow direction?

One of the most hotly contested matters among scientists is the river's age and thus that of the Grand Canyon. Some geologists see evidence for an old river and canyon, on the order of 80 or 70 million years. Others believe that they are quite a bit younger, between about 6 and 5 million years. However, upstream in the state of Colorado, the river shows evidence of being somewhere between 20 and 10 million years. How can a river be 20 million years old in one location but no more than 6 million years downstream? Novel ideas

The Vermilion Cliffs near Lees Ferry expose strata believed to have once covered the Grand Canyon area. Photograph by Gary Ladd

Aire Bueno
Acrylic on board
by Bruce A. Aiken,
1983

on how rivers can change and evolve through time were the result of this confusing set of circumstances regarding Grand Canyon. We will examine these ideas more closely and they include the suggestion that the Colorado River may have formed from the prior existence of multiple rivers that were integrated into the single river system we find today.

Lastly, the severe depth of the Grand Canyon in relation to the country that encloses it is also a puzzle. Broad, near featureless plateaus surround the canyon. Early travelers, arriving on horseback or stagecoach, were just as impressed with the remoteness and seemingly endless plateaus that delivered them to the canyon's edge as they were with the color and depth of the great gorge. It may not be readily apparent to the non-geologist that these flat, highly elevated plateaus are worthy of discussion but it is likely that they formed at a different time under different erosional processes than the deep canyons that dissect them. What sequence of geologic events could have produced such a strikingly different set of landforms so close to one another?

These puzzling relationships: flow through an elevated plateau, the lack of fault control on the placement of the river, possible reversed drainage direction, uncertain age, and the canyon's setting within a more mature landscape, help us to frame the questions we need to ask in order to understand the canyon's origin. There are many possible explanations for these questions and each explanation may raise more queries than it answers. And although it may seem frustrating at first to work on such a problem, we find that as geologists dig deeper into these mysteries their enthusiasm and determination seems to increase as they try to resolve the origins of this world-class landform.

Those curious enough to ask these questions rely on the scientific method to find a satisfactory answer. Using this method a careful observer will frame a question regarding a specific problem, for example, "How did the Grand Canyon form?" This observer will then propose a theory that might offer a possible solution to the problem. A new theory must not disregard other theories that previously addressed the same problem. It is then tested by the observer and unbiased colleagues and in time it may be proven to hold up better to scientific scrutiny than others. It may eventually come to be regarded as a valid answer to the original question. This scientific method has served us well in deciphering the complex geologic history of our planet.

Sometimes however, as is the case with the Grand Canyon, there is not enough evidence to bring everyone to agreement on a single solution. Because of this, there may be many possible answers and professional disagreements may ensue among geologists. Some of the most colorful and important discussions concerning the Colorado River's history have resulted in a "scientific draw" with no one answer fully accepted as truth. As interested observers, we must satisfy ourselves with the knowledge that we may never be able to fully explain what we see so vividly laid out before our eyes. In the end, this is the larger enigma of the Grand Canyon—that a feature so large and highly regarded may forever remain unknowable to us.

SUMMARY

The Grand Canyon is an enigma, a puzzle, a mystery. It is located in an unlikely location and follows a path that is confusing at best. Try as we might, we may never find a definitive answer to its origin and we may never know everything about it.

2

Physical Setting

The Grand Canyon is carved into a series of highly elevated plateaus in northern Arizona. Measured along the river the canyon is 277 miles long and contains numerous subdivisions within it. It is part of the great Colorado Plateau geologic province, and five factors have acted in concert through time to produce this stunning landscape.

KNOWING HOW the Grand Canyon formed requires an initial understanding of the larger landscape through which it is carved. When a person first imagines the canyon or sees a postcard of it, they might assume that such a remarkable feature is located among other nearby wonders such as the spectacular Rocky Mountains or Arizona's rugged Sonoran Desert. However, the Grand Canyon is not found near these other western landmarks, rather, it is carved into a series of flat, elevated plateaus that, to the visitor, may appear at first to be monotonous or uninteresting.

As one approaches the canyon from the south or north there is no hint as to what lies ahead. Some may even wonder if they are traveling in the right direction! Although the observant traveler will notice a few quick glimpses of the North Rim when approaching from the south on Arizona

Opposite: The Little Colorado River (lower right) enters the Colorado River below Cape Solitude at the south end of Marble Canyon. Photograph by Chuck Lawsen

Highway 64, most visitors rely on highway mileage signs as evidence they are on the right road. The numerous volcanoes seen just outside of Flagstaff or Williams, Arizona, may provide some visual relief for those who travel with only their destination in mind. But at Valle, a mere thirty miles south of the canyon's edge, the road drops down into a treeless flatland of grass and sage. Whether approaching the canyon from the north or south, one must travel across broad, seemingly limitless plateaus.

These plateaus are impressive landscape elements in their own right. As one prominent student of the canyon observed, if the state of Montana had not already co-opted the term, the plateau landscape surrounding Grand Canyon could justifiably be called "Big Sky Country." Oftentimes, spectacular cumulus clouds may be seen trailing away in all directions above the plateaus, their flat bottoms tracing out the gentle curvature of the earth. The sheer size of the plateaus immediately surrounding the Grand Canyon is impressive; collectively they are larger than the states of Maryland and New Jersey combined or the individual countries of Costa Rica, the Netherlands, or Switzerland. Their elevation varies with geologic structure from a low of about five thousand feet to over nine thousand feet above sea level. That means that the lowest point on these plateaus is just about the same elevation of the mile-high city of Denver, Colorado.

On the north side of the river we find four plateaus. From east to west they are the Kaibab, Kanab, Uinkaret, and Shivwits plateaus. On the south side are the Coconino and Hualapai plateaus. The upper stretch of the Colorado River in Marble Canyon bisects the Marble Platform, which is therefore found on either side of the river in that location. In all there are seven plateaus, each having its own distinctive character and story to tell. What separates the plateaus from one another are the various faults and folds that have cracked and warped the landscape like an old phonograph record sitting in the sun. As we shall see, some of the boundaries that separate the plateaus are defined by active faults which have played a major role in determining how the Colorado River has deepened the Grand Canyon.

In the eastern Grand Canyon, which is the part that most visitors see, the Colorado River separates the Kaibab Plateau to the

The Grand Canyon has been carved into a series of flat, nearly featureless plateaus, contrasting two very different landscapes and perhaps suggesting two different times and styles of erosion. Photograph by Wayne Ranney

north from the Coconino Plateau to the south. These plateaus were once a continuous feature before they were separated by the carving of the canyon. Both are part of the Kaibab upwarp, an elongate "blister" or dome on the earth's crust that was uplifted higher than the surrounding terrain. When the Colorado River cut the Grand Canyon, it did so not on the crest of the upwarp but on its southern flank, where the rock strata dip to the south at about one to two degrees. This gentle angle is not easily perceived by the human eye and some people are surprised to learn that the North Rim (at eighty-two hundred feet) sits about twelve hundred feet higher in elevation than the South Rim (elevation about seven thousand feet).

This southward slope causes more runoff to enter the canyon from the north side, which allows for more erosion in the side streams on the north side of the river. For this reason, the North Rim is eroded back away from the river about twice as far as the South Rim. Between the hotels on either side, the Grand Canyon is ten miles wide and one mile deep. It has been estimated that the amount of rock excavated from the Grand Canyon is just over eight hundred cubic miles a phenomenal amount of material considering that all the rivers of the world total only three hundred cubic miles of water! People often ask where all of that dirt has gone; the answer is that farmers are growing carrots and lettuce in it in the Imperial

The Grand Canyon *of the* Colorado River

The Colorado River winds 277 miles through the Grand Canyon from Lees Ferry (top right) to the Grand Wash Cliffs (center left), cutting through high plateaus divided by tributary streams.

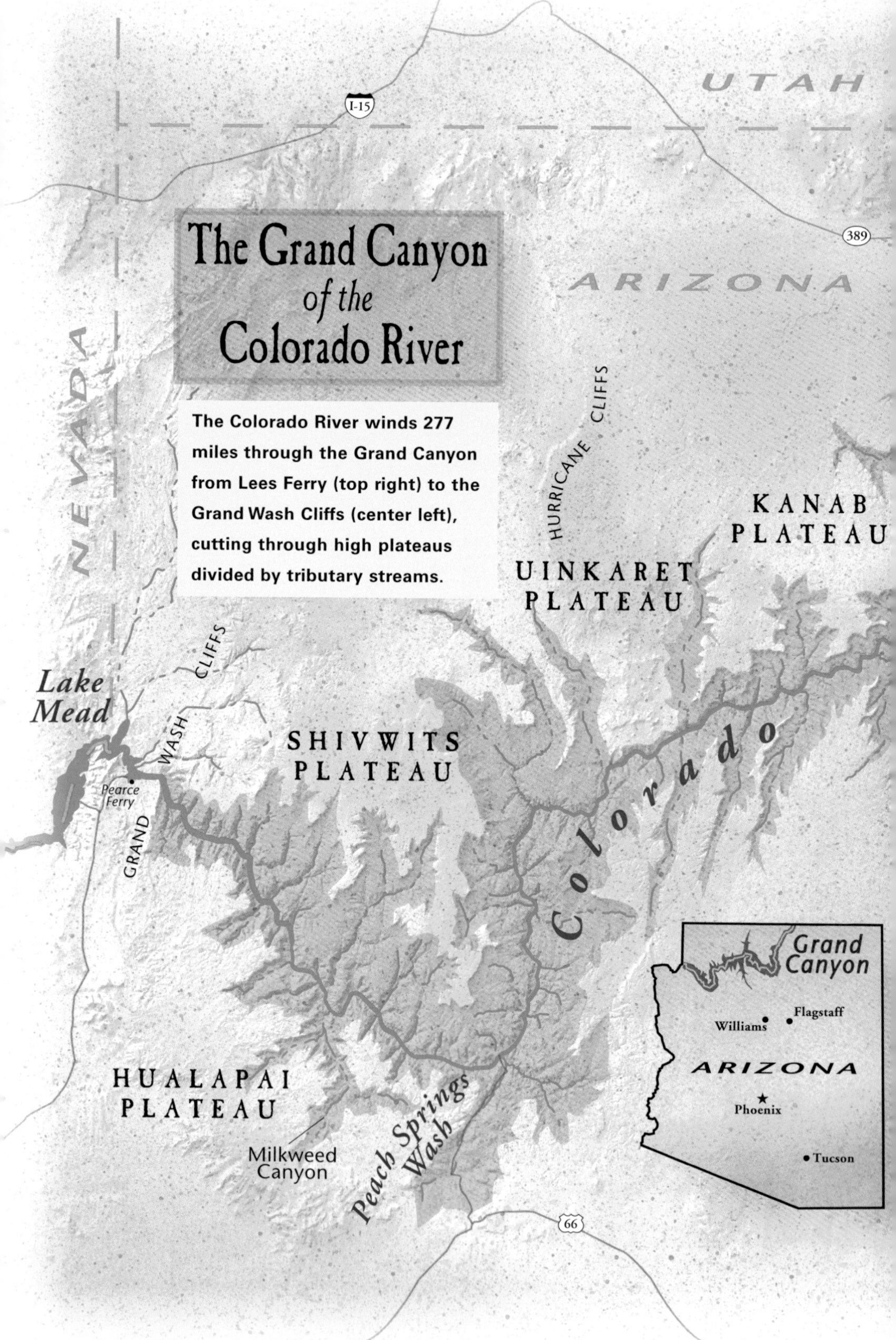

UTAH
I-15
389
ARIZONA
NEVADA
HURRICANE CLIFFS
KANAB PLATEAU
UINKARET PLATEAU
Lake Mead
GRAND WASH CLIFFS
SHIVWITS PLATEAU
Pearce Ferry
Colorado
HUALAPAI PLATEAU
Milkweed Canyon
Peach Springs Wash
66
Grand Canyon
Williams
Flagstaff
ARIZONA
Phoenix
Tucson

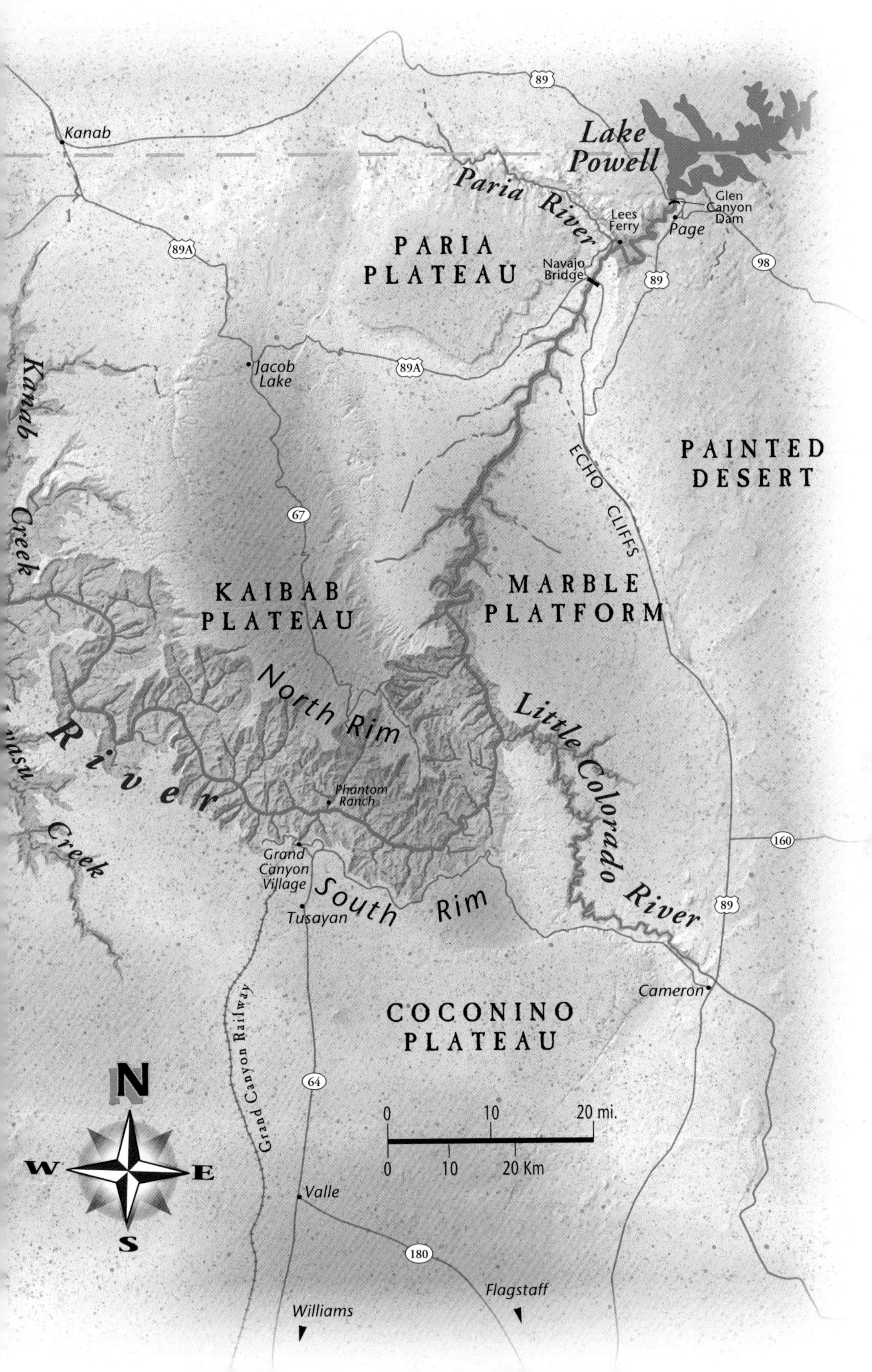

89
Kanab
Lake Powell
Paria River
Glen Canyon Dam
Lees Ferry
Page
89A
PARIA PLATEAU
Navajo Bridge
98
89
Kanab Creek
Jacob Lake
89A
ECHO CLIFFS
PAINTED DESERT
67
KAIBAB PLATEAU
MARBLE PLATFORM
North Rim
Little Colorado River
River
Phantom Ranch
Creek
160
Grand Canyon Village
South Rim
Tusayan
89
Cameron
Grand Canyon Railway
COCONINO PLATEAU
N
64
0
10
20 mi.
W
E
0
10
20 Km
Valle
S
180
Flagstaff
Williams

Valley west of Yuma, Arizona, which used to be where the Colorado River emptied into the Gulf of California.

Within Grand Canyon are a series of differentiated landforms as well. Located three quarters of the way into in the canyon is a broad, greenish terrace called the Tonto Platform. It has formed where the soft and easily eroded Bright Angel Shale has retreated away from the river at a much faster rate than the underlying sandstone and crystalline rocks. Beneath the Tonto Platform is the Granite Gorge, cut into relatively hard granite and schist. More than one thousand feet deep in many places, this dark, forbidding slot is the most recent part of the landscape to be revealed by the cutting of the Colorado River. Within the 277-mile length of Grand Canyon are three distinct reaches of the Granite Gorge—Upper, Middle, and Lower. It is the Upper Granite Gorge that is present beneath the hotels on both rims. Lastly, and seen only in the western two thirds of Grand Canyon is the vermilion-tinged Esplanade. Many aficionados of the canyon consider this landform to be the most graceful example of canyon architecture. It has formed where the soft Hermit Formation has retreated away from the river at a more rapid pace compared with more resistant formations below it.

Deep within the canyon, recession of the soft Bright Angel Shale has formed the Tonto Platform, seen in the bottom third of this photograph. The river slices from right to left through the canyon's oldest rocks to form the Inner Gorge. Photograph by Chuck Lawsen

The Esplanade in western Grand Canyon is rarely seen by most visitors. It likely formed by the retreat of the easily eroded Hermit Formation. Photograph by Gary Ladd

Located in the northern part of the state of Arizona (and not the state of Colorado as some people mistakenly believe), the Grand Canyon sits isolated from most popular destinations in the Southwest; Las Vegas and Lake Powell are the closest two and are located at either end of it. When visiting the canyon, travelers must invest a significant amount of time driving or flying over flat terrain that is deemed monotonous or even boring by some. And yet, it is precisely because the canyon is deeply set within these plateaus that it is so intriguing. Plateaus oftentimes represent mature landscapes but the extreme depth of the Grand Canyon suggests a more immature landscape that could be described as "under construction," still in the making. When one stands on the edge of the Grand Canyon and looks first forward and then back, two vastly different landscapes are seen, one very much vertical and the other very much horizontal. Each tells a separate part of the story and combined they tell us how the Grand Canyon regional landscape was formed.

The plateaus, then, are not merely something to quickly put behind us so that we may attain the prize that lies ahead. In fact, the plateaus give definition to the prize itself and together they combine to form a landscape that is truly unique upon our planet. The plateaus have, in a sense, surrendered any overt beauty they may contain to the obvious glories that lie within the walls of Grand Canyon. Many come to the Grand Canyon, a few as landscape pilgrims, to gaze deeply into this chromatic shrine. But upon arrival they should not forget the stupendous platform that has delivered them to the altar. In deciphering the story of how the canyon came to be, the plateaus set the stage for how the canyon became so grand!

The seven plateaus surrounding the Grand Canyon are part of a larger physiographic province called the Colorado Plateau. The United States is divided into approximately twenty-five different geographic provinces defined mainly by their unique landscape characteristics. The Piedmont, Appalachians, Great Plains, and Sierra Nevada are examples of other provinces located in our country. The Colorado Plateau is defined as a region of uplifted, but relatively flat-lying sedimentary rocks centered among the Four Corners states of Arizona, New Mexico, Colorado, and Utah. It contains colorful, elevated plateaus which have largely escaped the extreme mountain building processes, such as faulting, folding, and volcanism that have greatly influenced the development of adjacent provinces. Certainly, these three processes have had some effect on the development of the plateau landscape but not to the extent or magnitude of bordering areas. The Grand Canyon region is located along the southwestern edge of the Colorado Plateau. In fact, the spectacular and abrupt Grand Wash Cliffs define both the western edge of Grand Canyon and that of the Colorado Plateau.

The adjacent Rocky Mountain Province is located to the north and east of the Colorado Plateau. Rocks in this province are varied, being of sedimentary, metamorphic, or igneous origin but generally elevated much higher than the plateau. To the south and west of the plateau is the Basin and Range Province. This relatively young province formed when our earth's crust was stretched and thinned. Numerous basins have been down dropped and are separated by elevated ranges. Why the Colorado Plateau has escaped the intense deformation episodes that greatly altered rocks in the Rocky Mountains and the Basin and Range remains unresolved. Nevertheless, it is the graceful, flat-lying nature of the colorful sedimentary strata that gives the Plateau its special feel.

The Colorado Plateau has an average elevation of about six thousand feet and an extreme elevation of over twelve thousand feet. This means that precipitation comes mostly as snowfall in the winter

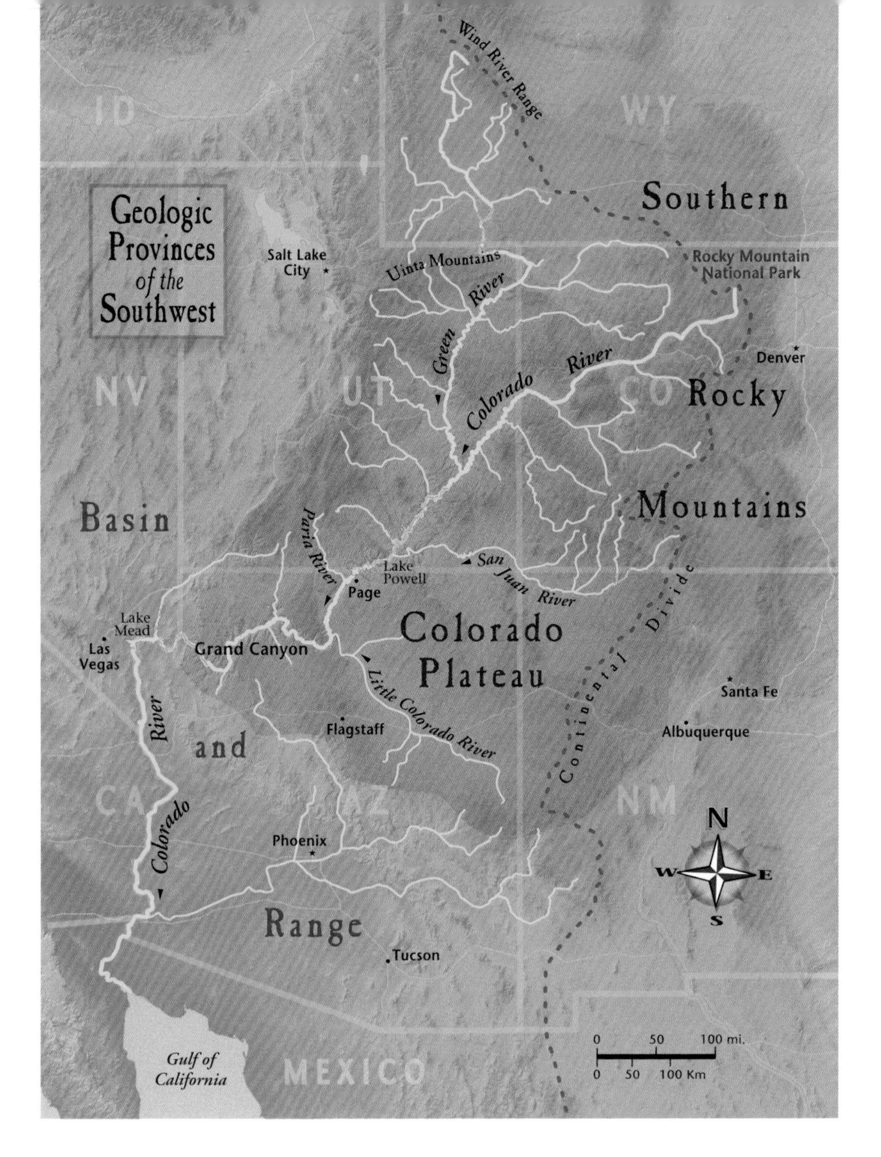

or infrequent but intense thunderstorms known as the summer "monsoon." The presence of bare rock exposures on the plateau, "slickrock" in local parlance, allows for catastrophic runoff that coalesces into the large rivers that originate in the Rocky Mountains. The Colorado River is the master stream which flows through the center of the plateau, giving the plateau its name. The river begins below the Continental Divide in Rocky Mountain National Park, Colorado, and in the Wind River Range, Wyoming, as the equally contributive Green River branch. Over 70 percent of the Colorado River's water originates in these headwater mountains. The major tributaries on the plateau are the Gunnison, Yampa, and Dolores

From headwaters to mouth the Colorado River flows through three distinct geologic provinces.

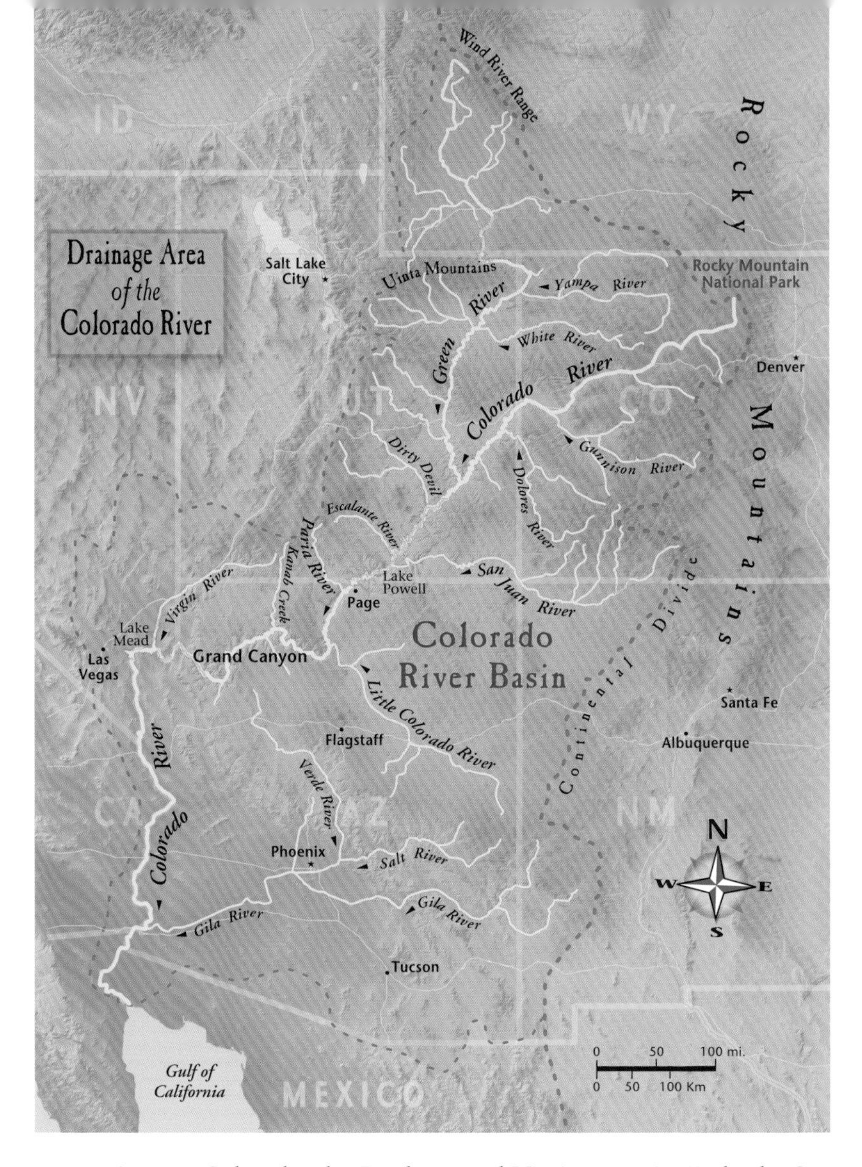

The Green and Colorado rivers gather water from numerous tributaries to drain the Colorado River Basin.

rivers in Colorado, the Escalante and Virgin rivers in Utah, the San Juan in Colorado, New Mexico, and Utah, and the Little Colorado River in Arizona. The plateau itself may be arid but it is coursed by rivers whose branches reach far back into the high backbone of our continent where precipitation is much greater.

Five rather independent conditions have acted in concert to produce the unique landscape of the Colorado Plateau. If only one of these were to be removed from the landscape history, the plateau as we know it today would not exist. There are a few places on our planet, such as central Asia, the Middle East, or central South America that approximate the look and feel of the Colorado Plateau but they are not as extensive in size nor do they contain the same combination of these conditions. These conditions are:

1) The presence of a thick stack of stratified rock
2) Vivid and varied color within these strata
3) The massive and widespread uplift of this strata so that it remains relatively flat-lying
4) An arid climate
5) The presence of large rivers flowing through the region.

One is immediately struck by the odd combination of an arid climate and the existence of large rivers. However, the Colorado River has its headwaters high in the Rocky Mountains of Colorado and Wyoming where precipitation is plentiful most years. The resulting runoff is channeled across the arid Colorado Plateau. There are only a few other places on our planet that combine these seemingly mutually exclusive conditions. Egypt and its Nile River is one of them, along with Iraq, which has the Tigris and Euphrates rivers. However, both of these areas lack one of the other conditions necessary to create another Colorado Plateau, uplift of the earth's crust upon which those rivers could carve a deep canyon.

The Colorado Plateau remains unique as a place where these five independent conditions have acted in unison to create this remarkable, rocky paradise. When one takes an imaginary journey from the lofty Continental Divide in the Rocky Mountains, down into the heart of the canyonlands and through the Grand Canyon to the Mojave Desert and the Gulf of California, he or she passes through one of the most interesting and beautiful transects of landscape scenery found anywhere on earth! This is the stage upon which the story of the Grand Canyon is set.

SUMMARY

The Grand Canyon is located in northern Arizona and has been carved upon a series of highly elevated plateaus. These plateaus are part of the larger Colorado Plateau, which is a unique landscape region on our planet.

3

How Rivers Carve Canyons

Rivers carve canyons when extreme floods roll large boulders along the bed of the river, causing the bedrock channel to become abraded and deepened. Coincident with these momentary events, but just as important, are the long-term changes in elevation, river gradient, and climate, as well as the growth and removal of newly formed obstacles.

In our rapidly urbanizing world, most people live their entire lives without ever thinking how rivers like the Colorado carve their canyons. However, upon arrival at the edge of the abyss that is Grand Canyon, their interest is piqued and many wonder how the river actually accomplished its task. Many of these visitors do not have a background in science or geology, yet the sight of the canyon is so stupendous and spellbinding that they naturally develop a curiosity about how its profound depth and beauty were attained. Oftentimes they conjure up ideas that imply a quick or catastrophic origin for the canyon, since it looks to some people like a raw, unfinished work of nature.

It is perhaps natural to think in terms of a catastrophic origin for the Grand Canyon since it does appear like a vast ruin, formed quickly and without design. Yet for many

The Inner Gorge is a spectacular example of canyon cutting on a grand scale. Photograph by Gary Ladd years within the park the simplified scientific explanation was that it was probably formed by the slow and inexorable wearing away of the bedrock by the vast amounts of silt and sand that once traveled down the river. Those vast amounts of sediment that once traveled through the Grand Canyon were trapped behind Glen Canyon Dam, located one hundred miles upstream from the main visitor area beginning in 1963.

When I was in grade school I heard that the Colorado River, in all of its brown, silt-laden glory, had slowly and imperceptibly cut through the rock over a long period of geologic time, acting somewhat like sandpaper working very slowly on a piece of wood. The teacher was using the most common explanation for how a gorge so deep could have formed. In my mind, I imagined the tiny particles of silt and sand running over the river channel, polishing the rock away bit by bit. In this manner one could live a couple dozen lifetimes and never notice the canyon getting much deeper.

It turns out, however, that very little deepening can be accomplished in this way. For one thing, when studies were carried out to determine what the bed of the Colorado River looked like under all that muddy water, it was found that the bedrock, defined as the solid, uneroded substrate beneath the river, is rarely exposed to the flow of muddy water. Up to seventy-five feet of silt, sand, gravel, and boulders deposited during floods, rested on top of the bedrock beneath the water. It's as if there is a protective coating of material that actually prevents the gritty river (the sandpaper) from wearing away the bedrock (the wood). When this was discovered, it became clear that the slow work of muddy water alone could not really have cut the Grand Canyon.

Continued observations of modern erosional processes suggested that rivers like the Colorado actually deepen their channels only during relatively rare and intermittent large-scale floods, when huge amounts of large, rocky debris are in motion. Flowing water has the curious property that when its volume during a flood is doubled, its ability to transport larger-size material is quadrupled. Thus, if the volume of water goes up by a factor of four, the size of the material transported goes up by a factor of sixteen; it's a squared relationship.

This is called a river's carrying capacity and the more water there is in a flood, the larger the carrying capacity of a river.

During huge floods, the carrying capacity of a river may be so large that the entire mantle of gravel and sand sitting on the bedrock is carried away downstream by the rushing flood waters. This exposes the bedrock channel to the flood waters and all that it carries with it. During these floods large, car size boulders may be rolling forcefully along the channel floor. As these boulders move downstream they physically pound the bedrock surface, breaking off huge chunks of it. In this manner, the bedrock channel can eventually be deepened by the intense physical abrasion of these large boulders. A single major flood could reveal some noticeable evidence of deepening, and four, six, or ten of these floods would eventually result in a channel that may have been deepened considerably. The waning stages of each flood typically leave a new mantle of sand and gravel, which puts the bedrock channel once again into a state of silent preservation beneath the muddy waters. Short-lived,

Left behind as floodwaters receded, these boulders in Marble Canyon represent the type and size of materials that grind into the bedrock channel under flood conditions. Photograph by Gary Ladd

intense flood episodes are responsible for creating the great depth of the Grand Canyon and these flood episodes may have been more frequent in times past when the climate was different.

Other factors are at play as well. This is illustrated in an encounter I once had with a visitor at the South Rim. I had just finished a short talk about the origin of the Grand Canyon to a tour group and I was staring deeply into the canyon at Yavapai Point. A member of our group approached me and asked, "Why don't we have a Grand Canyon in Minnesota?" The question seemed odd to me at first, for as a native westerner I assumed that everyone accepted the fact that deeply dissected landscapes were found in the West and that subdued, mature landscapes were the norm in the Midwest. I had never before considered why there wasn't a Grand Canyon in Minnesota. I soon realized that his question relates to one of the most important conditions that must be satisfied for the Colorado River to carve the Grand Canyon.

The Mississippi River runs through the Twin Cities of Minneapolis and St. Paul and is the dominant landscape feature in that large urban area. In my talk I had mentioned that the Mississippi River carried, on average, ten times the volume of water carried by the Colorado, trying to illustrate the point that as far as North American rivers were concerned the Colorado was not really a big river. This man had astutely surmised that if the Mississippi was ten times the size of the Colorado, then perhaps there should be a great canyon in Minnesota (maybe even ten times bigger). For a moment I was troubled by this line of reasoning but then I realized that I had failed to mention the most important process needed in the creation of the Grand Canyon—uplift. Without the gradual uplift of the landscape rivers cannot cut down into it.

When the light went on for me that sunny day at Yavapai Point, I turned to this man and responded with another question. "What is the elevation of the Mississippi River in Minneapolis?" We both agreed that it was about seven hundred feet above sea level. I then asked him, "How could a one-mile-deep canyon be cut into a landscape that is only seven hundred feet above the sea level?" I then watched as a light came on for him as he understood that this would be quite impossible. Uplift or elevation of the land is necessary to carve canyons and without it there would be no Grand Canyon.

Uplift of a landscape can be a difficult concept to grasp because in the course of our short lifespans, the earth appears to be stable and unchanging. But remember that geologic processes, although appearing quite slow to human beings, are forever present in our daily lives. At this very moment, the western side of North America is under pressure as it drifts southwest relative to the edge of another piece of the earth's crust, the Pacific Plate. The result is that the western part of our continent is being crumpled and uplifted, much like a throw rug (North America) when pushed along a hardwood floor into a wall (the Pacific Plate). These lateral pressures, which have been at work in western North America for many millions of years, are responsible for some of the vertical uplift of the southwestern landscape.

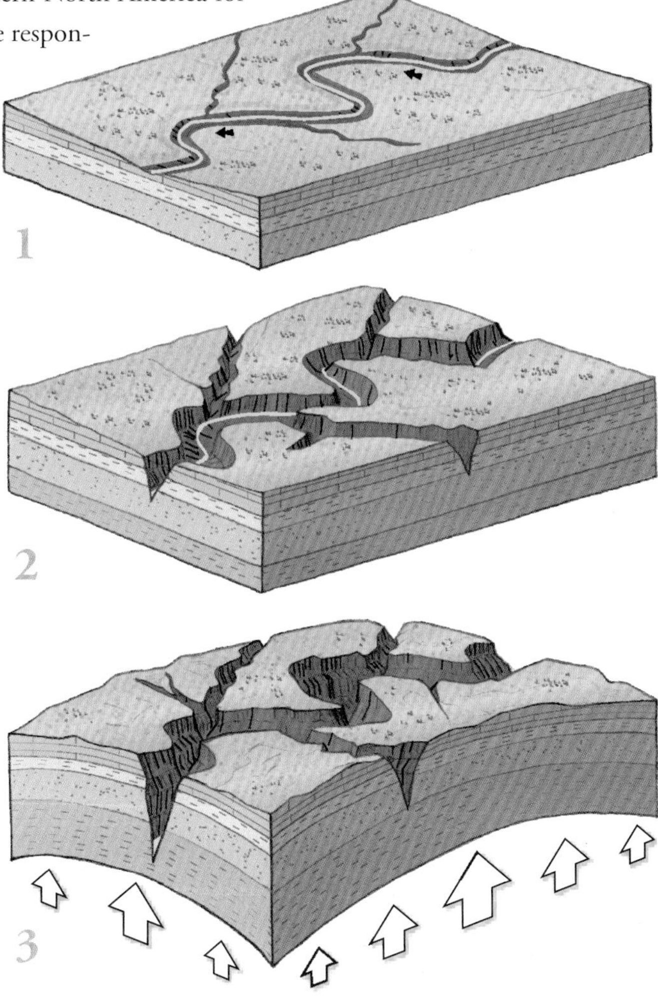

Progressive uplift leads to downcutting.

1) A river flows sluggishly on subdued terrain.

2) As the landscape is uplifted, the river incises accordingly.

3) Differential uplift raises the center part of the landscape at a faster rate than the edges. Canyon cutting keeps pace with the rate and amount of uplift.

Other forces may be responsible for uplift as well. Heat is constantly being generated within the earth and as it escapes, it causes the crust to rise like a hot air balloon rising through the cool morning air. Another idea is that the crust beneath the Colorado Plateau was perhaps once firmly attached to the underlying mantle which is quite dense. Perhaps the plateau was previously being "held down" by attachment to the mantle. But if the two became "delaminated" from each other, it would cause the crust of the plateau to rise up as if a weight (the mantle) was released from a piece of Styrofoam floating in water (the crust). Uplift can also occur when erosion removes the confining weight of overlying rocks. This type of uplift is called isostatic uplift. The exact reason why uplift has occurred in the Grand Canyon region remains speculative but certainly the area has been significantly elevated since the sea last left the area about 80 million years ago. This uplift has gradually raised the underlying strata into a position where flood waters can attack it. Water is always working to excavate down to sea level and during uplift it will work "harder" to achieve it.

As we shall see, the specific timing of plateau uplift is one of the most critical constraints on determining the precise age of erosion for the Grand Canyon. While it can be relatively easy to date a volcanic rock, the dating of an uplift is much more difficult to discern. As we examine the controversy on the age of the river and the canyon, we will constantly be referred back to the question of when the uplift of the Colorado Plateau occurred. Some geologists think that three distinct periods of uplift occurred throughout the last 70 million years. Some see the uplift early in this time period as being dominant, others argue for later uplift as being most important. Still others argue that it is the *differential uplift* that is more important, whereby portions of the plateau are found to be at a higher elevation due to the *lowering* of adjacent areas. There is still no consensus on when the uplift of the Colorado Plateau occurred.

However, recognition of recent movements on two faults in the western part of Grand Canyon are providing insights into how differential uplift could have significantly deepened the eastern part of the Grand Canyon in just the last few million years. The Toroweap and Hurricane faults cross the Colorado River at right angles, and

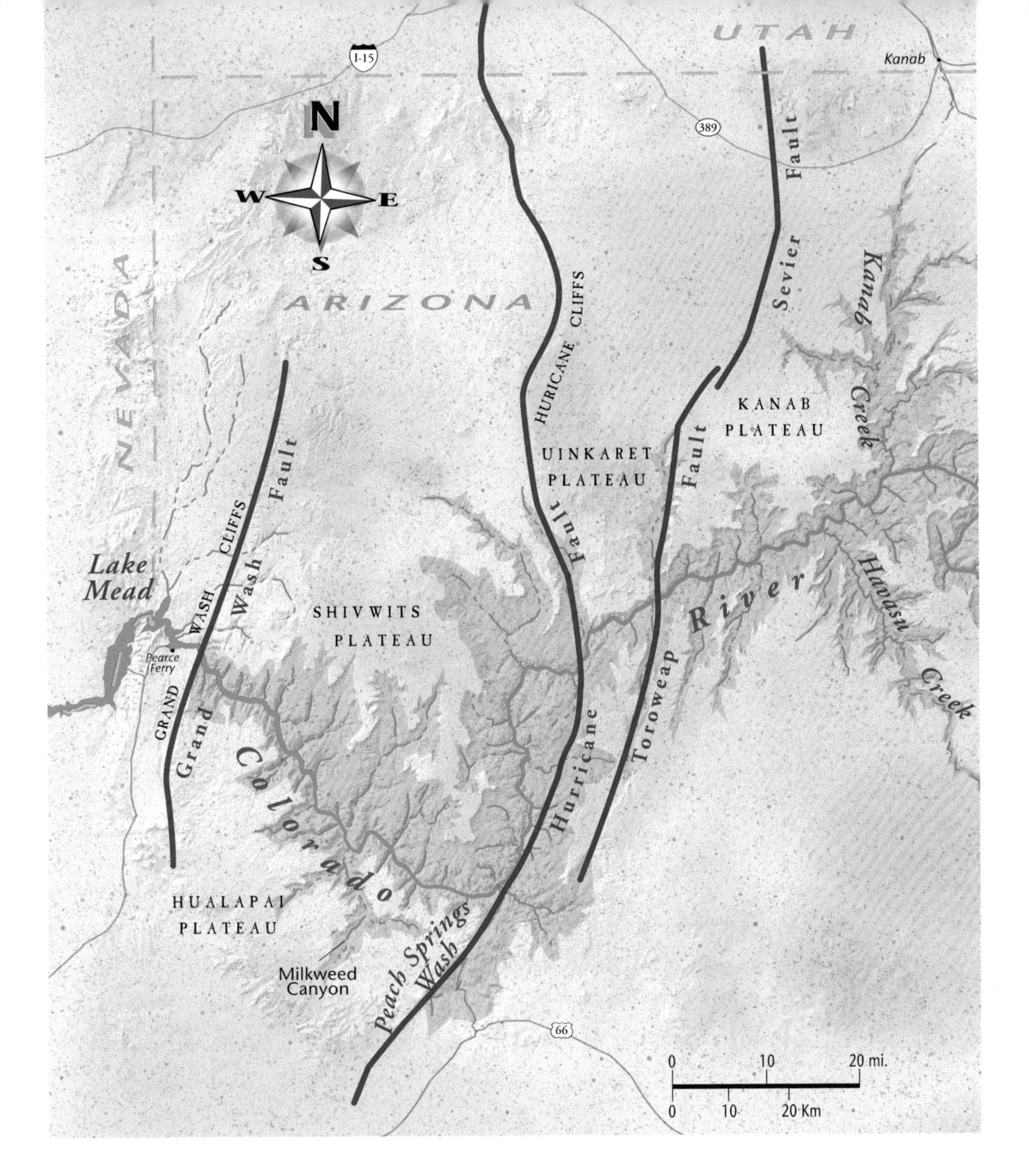

in both instances these faults raise the land on the east, or upstream, side. This results in the uplift of the river channel and the margin of this uplift is called a knickpoint. Knickpoints thus formed are prone to erosion and consequently, as the knickpoint is attacked by flowing water, it migrates upstream. This upstream migration of a knickpoint can cause a canyon to become deepened in that direction.

Niagara Falls is the perfect example of knickpoint migration. Before it was altered to produce electricity, the lip of the falls migrated upstream almost four feet per year—an astonishingly fast

Over the past 3.5 million years, movement along the Hurricane and Toroweap faults has differentially uplifted lands to the east 1,900 feet while the river's cutting action kept pace.

Niagara Falls is perhaps the most recognizable knick-point on the North American continent. Photograph by Jack Hillers, 1886

rate even in a human lifetime! The Niagara Gorge downstream from the falls is now seven miles long and has been chiseled down about one hundred sixty feet—all in just the last ten thousand years. Although the creation of this knickpoint is not the result of differential uplift along a fault, it is a good example of knickpoint migration that deepens a river canyon in the upstream direction.

In the Grand Canyon, knickpoint migration works in much the same way as Niagara Falls except that knickpoints in the canyon are created by differential uplift along the Toroweap and Hurricane Faults. The cumulative uplift along these faults during the last 3.5 million years is about nineteen hundred feet. This means that although the western portion of the canyon has been downfaulted (and thus perhaps not deepened significantly during this time), two fifths of the depth of the eastern part of Grand Canyon was probably accomplished in this short time. As we can now see, relative uplift is a very important concept in understanding the story of how the Grand Canyon has evolved.

Without relative uplift, it is not easy for a river to cut deeply into bedrock and the history and timing of plateau uplift is one of the keys to understanding Grand Canyon's origin. Yet climate must also play an important role since it determines the amount of runoff (and thus the carrying capacity of water) that can attack an actively uplifting terrain. Uplift may bring the rocks up to levels where they can be attacked but it is the elements of weather and climate that do the attacking.

An ancient debris flow deposited nearly seventy-five feet of sediment along Bright Angel Creek. Photograph by Wayne Ranney

We have seen how increased runoff can deepen canyons but what about the widening of canyons? Climate, specifically freeze and thaw processes, help to break apart rocks located far from the river channel. When water gets into cracks, it can freeze and expand, prying the cracks open. Then gravity takes over, pulling chunks down slope and eventually widening the canyon. If heavy rain saturates the landscape, whole hillsides can fail into a watery slush that moves like wet concrete down canyon floors. These are called debris flows and they work to actively widen portions of the Grand Canyon. Wind probably does very little to physically wear away rocks in the canyon but it does transport material away that has been already broken into sand-size grains.

In the wetter parts of our globe increased precipitation and humidity is responsible for the chemical breakdown of rocks and generally a more subdued landscape is created. In these tropical settings the surrounding hills are lowered at about the same rate that rivers cut into their channels. However, because the Colorado Plateau is located in an arid climate, erosion proceeds more slowly in areas away from the river's edge. Faster erosion occurs within river channels where infrequent but intense storms coalesce severe runoff into these narrow corridors. In the American Southwest, there is presently much greater vertical dissection along river courses, while the larger intervening areas of plateaus and mesas are left standing relatively high and dry. About 55 million years ago the Colorado Plateau was far more humid than it is today and it was during this time that the broad plateau surfaces may have developed.

Understanding climate change through time allows us to better understand how the different parts of the present landscape may have evolved. When the sea last retreated from the Grand Canyon region some 80 million years ago, the American Southwest was located in a humid, sub-tropical climate belt. The thermal maximum, defined as the hottest and most humid time, occurred about 55 million years ago. Erosion at that time must have been much different than what we see today. Broad, planar erosion most likely removed thick sheets of sedimentary strata that used to sit upon the plateau surface above Grand Canyon. These colorful strata are still present to the north in Zion and Bryce national parks but may have been removed from the Grand Canyon region during that humid time period. This may be when the Grand Staircase developed on the Colorado Plateau landscape. One of the great canyon geologists, Clarence Dutton, called this cycle of lateral erosion the "Great Denudation."

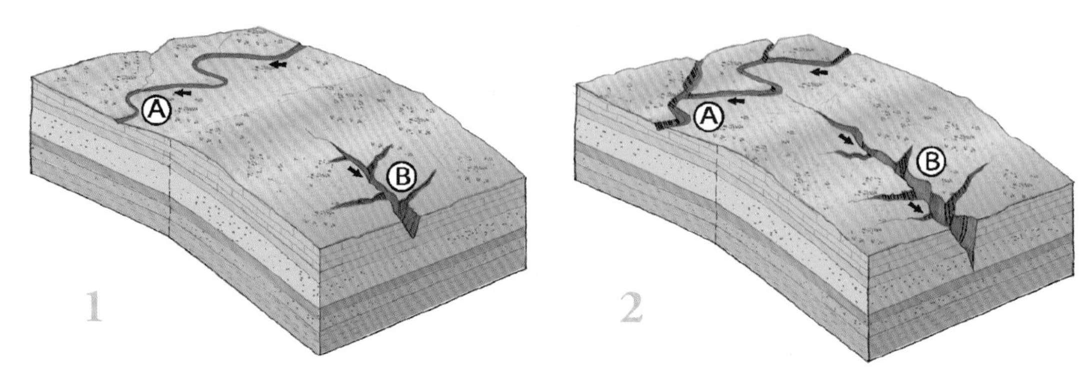

The related concepts of headward erosion and stream capture are important to understanding how rivers carve canyons. Portions of the Colorado River, both within and outside the Grand Canyon, may have formed in this way. 1) A sluggish river (A) and a steep-gradient river (B) share the landscape. 2) Because of its steeper gradient, B extends its channel into the slope toward A.

Slowly through time, as the climate began to cool and become dryer, landscape-wide planar erosion probably gave way to more localized erosion along river courses. Significant aridity in the southwest may have begun about 15 million years ago and increased in intensity between 5 and 6 million years ago. This more recent time period is when rivers on the plateau began cutting their deep canyons. Many recognizable aspects of the modern, highly incised landscape were most likely beginning to form during these more arid conditions. Uplift and climate change may have worked hand in hand to create the Grand Canyon landscape. Although the timing of this uplift remains unresolved, the broad aspects of the West's climate regime are becoming better known, allowing scientists to interpret how and when landscape development may have progressed in the absence of river-borne or uplift evidence.

The related concepts of headward erosion and stream piracy are also important in understanding how rivers carve canyons. As we shall see, there are conflicting dates for the age of the Colorado River both above and below the Grand Canyon. How can a river be old in one place but young in another? At first this may seem quite inexplicable but if the river evolved from two separate systems, only to be joined at some later time into an integrated single river, then the puzzle is easily solved. Stream piracy, accomplished by the process of headward erosion, provides us with a likely set of circumstances that can explain many enigmas concerning the Colorado River.

Stream piracy suggests that for two rivers, each located on opposite sides of a drainage divide, the steeper gradient stream will

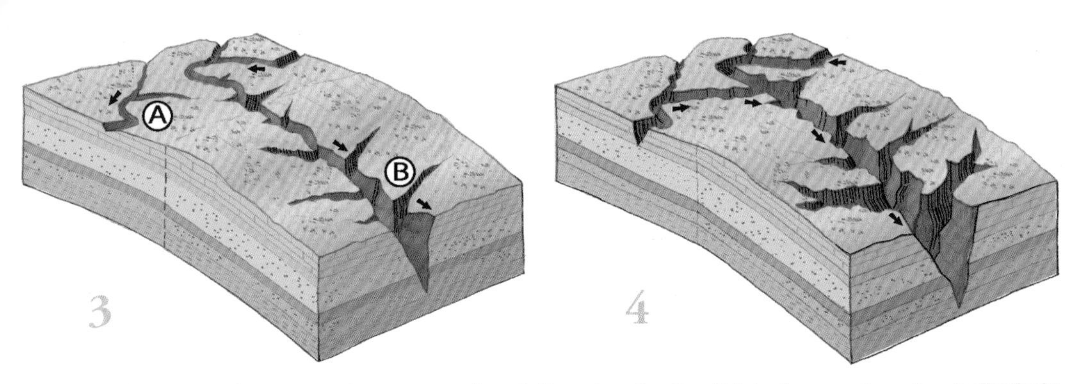

3) B eventually captures the upper portion of A's water, diverting it into its own channel and effectively "beheading" stream A. 4) Continued erosion eventually results in a fully integrated river system retaining the configuration of both streams and cutting deeply into the landscape.

preferentially lengthen its channel in the upstream direction, "capturing" terrain that previously belonged to the lower gradient stream. This happens because steeper gradient streams can move more and bigger material, causing this stream to more readily deepen its channel. This increased down-cutting eventually causes the stream to undercut and steepen the headwall area near the drainage divide.

As the drainage divide becomes undercut and steeper, it naturally wants to return to its more stable profile. The only way the divide can do this is to migrate farther back into the areas formerly occupied by the lower gradient stream—headward erosion in action! When this headward erosion finally reaches the channel of the lower-gradient stream, it will have integrated the upstream portions of the low-gradient stream as part of its own steeper stream system. Headward deepening can then proceed up this channel; we can think of the position of the lower gradient stream as a pre-determined "perforation" in the landscape that dictates where headward erosion and deepening will occur.

By definition then, the downstream reach of the low-gradient stream will have become "beheaded," since its upstream reach has been captured by the high-gradient stream. Eventually headward erosion may proceed along this pre-existing "perforation" as well, thus changing the direction of runoff. This reach of the stream will then be said to have undergone drainage reversal. All of this happens over what may seem a long time for humans, but it occurs quickly in geologic terms. Through time certain rivers can capture parts of other drainages, changing their overall configurations dramatically.

Headward Erosion Today

Headward erosion may be responsible for the integration of ancient separate river systems into the modern Colorado River we see today. Although it may be impossible to know with certainty if this process alone produced the course of the modern river, a question presents itself: Can evidence be found upon today's landscape where headward erosion may be occurring, leading to stream capture and the integration of distinct river systems, and thus adding more drainage area to the Colorado River? The answer may be found at the headwaters of the Paria River within Bryce Canyon National Park.

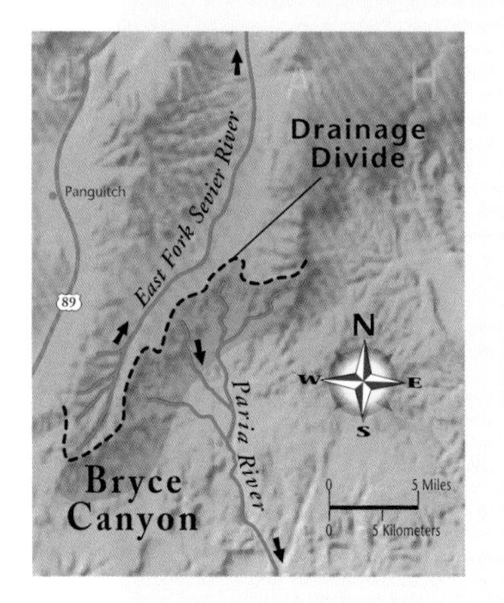

The rim of Bryce Canyon forms a drainage divide between two separate river systems—the Paria River on the east and the East Fork of the Sevier River on the west. A drop of rain water falling below the rim of Bryce Canyon will flow south into the Paria and on to the Colorado River system. But a drop of rain that falls on top of the rim will flow north into the Sevier River and the Great Basin. The East Fork of the Sevier has a very low gradient, suggesting that it is a mature stream which may be as old as 35 million years. The headwaters of the Paria River have steep gradients, indicating that they are young and vigorous. The colorful hoodoos at Bryce Canyon are formed as the Paria River's headwall erodes back into the soft and colorful Claron Formation, creating the Paria Amphitheater.

Just two miles from the rim of this amphitheater is the Sevier River. The rim of Bryce Canyon is cutting toward the Sevier River channel at the rate of about one foot in every sixty-five years—a phenomenally fast rate of change in geologic terms. A quick calculation indicates that continued retreat along this drainage divide would

It is possible that the Colorado River has experienced such episodes of headward erosion and stream piracy within the Grand Canyon.

One other interesting consideration in understanding how rivers carve canyons is to note the relationship between the perennial Colorado River and the many dry tributaries that enter it within Grand Canyon. Most of these tributary streams enter the river at grade, meaning at the same level as the main river channel. (There is one notable exception at Deer Creek Falls, but this exception is due to a relatively recent catastrophic change in the creek's position.) How is it that much smaller tributaries, which have no water cause the rim of Bryce Canyon to arrive at the Sevier River channel in about 686,400 years. Climate change, relative uplift, or changes in runoff amounts could affect the rate of change. But when the rim of Bryce Canyon reaches the Sevier's East Fork, all of the water upstream from the point of capture will then be directed east and south to the Colorado River system. As headward erosion progresses along the downstream section of the Sevier River, drainage reversal will occur on this stretch of the river.

This present-day landscape relationship lends credence to the idea that the Colorado River may have been cobbled together in times past from separate and distinct river systems.

Left: The steep-gradient headwaters of the Paria River are creeping toward the low-gradient Sevier River through the drainage divide. Above: The Paria Amphitheater at Bryce Canyon Photograph by Tom Till

in their channels most of the time, can carve canyons just as deep as the Colorado River has carved the Grand Canyon?

The answer to this puzzle is that most canyons are carved only during relatively rare flood events when large amounts of debris physically abrade stream channels. Even though the dry tributaries in Grand Canyon are inactive 95 percent of the time, their steeper gradients cause them to transport much larger material during floods, meaning that they can carve into bedrock just as readily as the much larger Colorado River has in the past. In fact, we could say that the tributaries have enough potential power to cut their canyons even

Many dry streams, like Pigeon Canyon (above), enter the Colorado River at grade. Photograph by Scott T. Smith

faster and deeper than the Colorado cuts its canyon, since their gradients are much steeper, but of course, they can only cut as deep as the master stream. In a very real way, the present-day mature gradient of the Colorado River is inhibiting the tributary canyons from becoming deeper!

Now that we have examined the various ways in which rivers can carve canyons, we can address some misconceptions that still exist regarding the Colorado River. When visitors are told that the canyon is ten miles across from El Tovar Hotel on the South Rim to the Grand Canyon Lodge on the North Rim, many may assume that the Colorado River must have been much wider at one time than it is today. In fact, the river has probably always been about as wide as it is now, about three hundred feet on average, throughout its time within the walls of Grand Canyon. During the Ice Age, when glaciers were repeatedly melting in the Rocky Mountains, the river's volume may have varied quite significantly but never was it even marginally close to being ten miles wide.

The canyon's great width is the result of tributaries to the Colorado River actively incising their channels, forcing the rim on

Tapeats Creek, a much smaller stream than the Colorado River, has cut its canyon just as deep as Grand Canyon. This suggests that it is the gradient of streams, and thus their carrying capacity, that determines the rate of canyon cutting. Photograph by Larry Lindahl

How are rivers placed upon a landscape?

Streams find their way across a landscape because water responds to the force of gravity by seeking the lowest path through varied topography. However, erosion has been working for at least 70 million years on the Colorado Plateau and this has sometimes left a confusing set of relationships, in which some rivers seem to flow towards or through areas with high topography. For these reasons, it is instructive to know some of the ways that rivers can be initially positioned on a landscape.

A stream that originates by flowing down a tilted surface is called a consequent stream. In opposition to this, a river that flows against the dip of stratified beds is called obsequent. Geologists assume that obsequent streams started out as consequent streams which later reversed their flow direction. A stream that is positioned in a strike valley, which is a valley formed along a fault line or a belt of easily eroded soft rock, is called a subsequent stream.

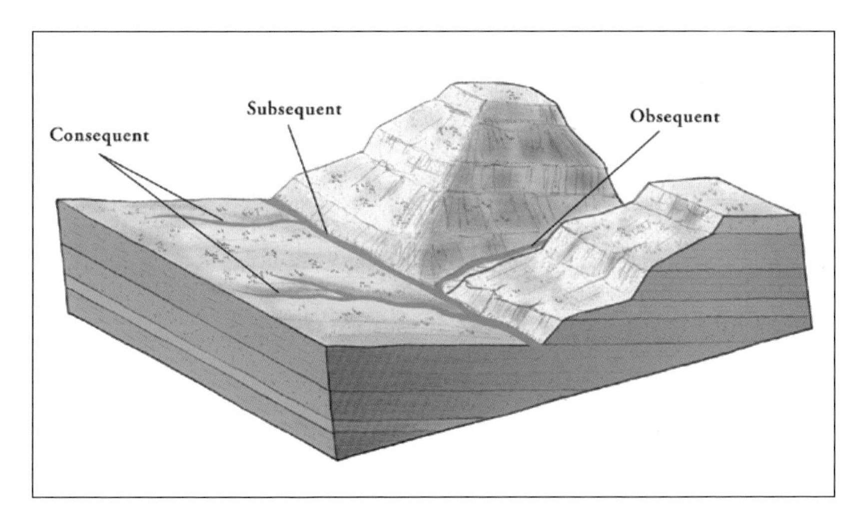

If a river originates on a subdued surface, say the flat bed of a dry lake, and later cuts down below this surface into deeply buried but irregular bedrock, it is termed a superposed river. A superposed river can eventually cut down into buried (but pre-existing) topography no matter how resistant rocks in that buried topography are. See illustration on page 66.

Alternatively, if a river begins on a subdued landscape that is differentially uplifted slowly around it, such that the river's course is not deflected away from the rising structures, it is termed antecedent. See illustration on page 65. The Colorado River system may display evidence for each of these variations in different segments of it's length.

either side to become undercut and retreat back away from the main river channel. Once the Colorado cuts into the rock layers, the tributaries work hard to keep pace with the deepening main gorge. All of these "open cuts" become susceptible to the other forces of erosion such as freezing and thawing, undercutting, and gravity that tend to wear away the edges of exposed rock surfaces. While it is true that the layers at the top of the canyon have been exposed to erosion for the longest time, probably more important is the faster rate at which underlying soft shales erode, undercutting and collapsing the harder layers that overlay them. This is most likely why the canyon is wider at the top. Differential erosion of the variable strata is the key to canyon widening.

The Colorado River has cut through about nineteen different sedimentary formations, as well as at least two kinds of crystalline rocks at the very bottom. Each formation has a certain resistance to erosion based on its rock type, hardness, or density. It's obvious that a granite or schist is much more resistant to erosion than a shale layer. Many people assume then, that when the river is cutting through granite its rate of down-cutting will be slower than when it cuts through a rock like shale. This has been a long-held assumption regarding canyon formation and it seems to make sense.

However, there are overriding phenomena that render this line of reasoning essentially meaningless. Canyon cutting may occur within discrete periods of time that involve the larger and more important parameters of active uplift, climate change, or increased runoff. When these conditions are active, a river such as the Colorado may not care one bit about the type of material it must saw through, at least with respect to deepening. In other words, it may not matter to any great degree whether the river is cutting through shale or granite—if uplift, climate change, or increased runoff are in play, the channel will become deeper, and rather quickly at that. Rock type does play a significant role in determining the overall width and profile of the canyon walls but other factors are more important in determining the depth.

These observations are relatively new and were made when geologists looked at the various lava flows that once blocked the Colorado's path. At least thirteen lava dams have existed within the canyon during the past 630,000 years and some of them were quite tall or long. Yet, these dams were destroyed, then replaced by subsequent lava dams within a relatively short period of time. This shows that no matter what type of obstacle was put in the river's path (and basalt lava can be quite hard and dense), it is removed relatively quickly regardless of its resistance to erosion. True, these lava dams may have been inherently unstable from the beginning since the hot lava was erupted on top of unconsolidated river gravels and/or was quenched quite quickly in cold river water; the dams may have collapsed catastrophically shortly after they formed. But some of these dams are documented to have been over 2,000 feet high and one was more than eighty-four miles long. Yet these obstacles were removed relatively quickly one way or another from the river channel before the next dam was emplaced.

As a geologist I am often asked about the effect that Glen Canyon Dam may have on the rate of deepening of the Grand Canyon. The line of reasoning is that the dam has tamed the wild river and its springtime floods, thus stopping the deepening which must have occurred prior to the dam's construction. Certainly, the

Glen Canyon Dam, located thirteen miles upstream from the head of the Grand Canyon, temporarily obstructs the river. Photograph by Stewart Aitchison

Storm over the North Rim near Bright Angel Point
Photograph by Bob and Suzanne Clemenz

dam does hold back any large floods and the sedimentary debris that would have previously roared through Grand Canyon. In the short term, Glen Canyon Dam has greatly affected the sediment load and the ecology along beaches within Grand Canyon.

However, with respect to geologic time, Glen Canyon Dam is a temporary feature and will be obsolete within a few hundred years. Even if we use the youngest age for Grand Canyon ascribed by some geologists, 6 million years, the life span of Glen Canyon Dam pales in comparison. We must remember that the Grand Canyon's evolution has most likely proceeded in fits and starts and that even prior to humans' meddling, there were significant periods of time when Grand Canyon was not actively being incised. These natural "breaks" in canyon cutting far surpass the time that Glen Canyon Dam will be in existence. Perhaps the river will not even notice that a barrier of such short duration as Glen Canyon Dam even existed. Glen Canyon Dam is such a temporary feature in geologic time that it has no effect on retarding the overall rate of cutting of the Grand Canyon.

And so, one of the pictures that is beginning to emerge is that the Colorado River may have cut the Grand Canyon in episodic pulses of active erosion and deepening, interspersed with long periods of quiescence and stability with regard to its depth and width. This concept is similar to that developed by evolutionary biologists for the descent of species, called "punctuated equilibrium." In other words, the greater portion of the canyon's history may involve periods of stability or equilibrium, where no deepening or widening of the canyon occurs, that are disrupted episodically by periods of great change due to increased uplift, climate change, or runoff amounts. This impacts our ability to say with certainty "how old" the Grand Canyon may be but it frees us to better understand why that question could have so many answers.

SUMMARY

Many factors are involved in the task of carving canyons. And if several conditions act together canyon cutting can occur quite rapidly, at least with respect to geologic time. Large-scale flooding, uplift, climate change, and headward erosion and stream piracy are some of the important elements that cause rivers to carve canyons.

4

History of Geologic Ideas

The Grand Canyon at sunset
Photograph by Chuck Lawsen

Theories on the origin of the Colorado River have evolved through time. Although one coherent theory about the specific details is still elusive, a broad, big picture understanding of the canyon's origins is emerging. However, some of the questions that originated in the nineteenth century remain unresolved in the twenty-first century.

WE HAVE EXAMINED the general ways in which rivers carve canyons. And as we now turn our attention to the evolution of ideas regarding Grand Canyon's origin, we see not only how scientific methods advance our understanding of the canyon but also how Grand Canyon has pushed geologists to better understand the very same processes that created it. How interesting it is that in trying to unravel the intricacies of the canyon, we have advanced our general understanding of how rivers respond to changing climates, uplifted terrain, and stream piracy. The Grand Canyon truly is a remarkable place.

This chapter examines the evolution of the myriad ideas put forth by geologists over time as they grapple to understand Grand Canyon's formation. The chapter is divided into four parts: nineteenth century, early twentieth century, late twentieth century, and the twenty-first century. Most of the major scientific ideas ever proposed are included, as well as the major themes that provoke debate. The date after the name of the geologist refers to the publication date of their ideas. Specific references for these and other theories are listed in the bibliography.

Late Nineteenth Century

John Strong Newberry, 1861

A common misconception among students of the Grand Canyon is that John Wesley Powell was the first geologist to view it and attempt to explain how it came to be. In fact, that honor goes to John Strong Newberry who accompanied Lieutenant Joseph Christmas Ives on his pioneering exploration of the lower Colorado River in 1858. Lt. Ives has been immortalized in the Grand Canyon region as the man who least understood the significance of the landscape he was instructed to explore. A quotation from his report to the Congress of the United States, often recalled today by Grand Canyon enthusiasts, bears repeating.

The region . . . is, of course, altogether valueless. It can be approached only from the south, and after entering it there is nothing to do but leave. Ours has been the first, and will doubtless be the last party of whites to visit this profitless locality. It seems intended by nature, that the Colorado River, along the greater portion of its lonely and majestic way, shall be forever unvisited and undisturbed.

He couldn't have been more wrong! Today over four and a half million people visit his "profitless locality" each year, 40 percent of them from beyond the borders of the United States. But rather than scoff at Lt. Ives's lack of vision (for in fairness to him he did pen some quite complimentary passages about the canyon landscape in the same report), it is helpful to understand the point of reference from which he was speaking. He hailed from New Hampshire and surely held the belief common in those days that a place as dry and highly dissected as the Grand Canyon could never amount to anything good for our young nation. Although his main objective was to determine whether the Colorado River could be used to supply military troops engaged in the Mormon campaign, he was likely predisposed to report on the presence or absence of arable land that could be settled and cultivated. Knowing his nineteenth-century bias makes it easier for us to understand why he wasn't impressed. It would take a different, less agrarian orientation to understand that perhaps the Big Cañon, as he called it, would have some redeeming value in the future.

Amazingly, that point of reference would come from within his own exploration party! Among his group was John Strong Newberry, who looked upon the rugged terrain from the same vantage points but saw a much different landscape than Ives did. He was trained both as a physician and a geologist and as such was prepared to accept with greater appreciation the spectacular natural feature that their party came upon. He wrote these words about the Colorado Plateau region shortly after one of his expeditions north of Grand Canyon:

Though valueless to the agriculturalist, dreaded and shunned by the emigrant, the miner, and even the adventurous trapper, the Colorado Plateau is to the geologist a paradise. [It has the] most splendid exposure of stratified rocks that there is in the world.

Professor Newberry was looking at the same landscape as Ives and struggled with him in their futile attempts to find reliable sources of water for themselves and their horses. However, he was capable of seeing that the landscape laid out before them was truly unique upon our planet. He was able to make the first elementary, but ultimately necessary observation that it was *erosion by running water* that had created this great chasm:

Having this question constantly in mind, and examining, with all possible care, the structure of the great cañon which we enter, I everywhere found evidence of the exclusive action of water in their formation. The opposite sides of the deepest chasm showed perfect correspondence of stratification, conforming to the general dip, and nowhere displacement; and the bottom rock, so often dry and bare, was perhaps deeply eroded, but continuous from side to side.

In this powerful passage, we see how a trained eye can observe the evidence, without preconception or bias, and make the most fundamental interpretation for what shaped the landscape. Newberry noticed that the sequence of strata was not displaced or broken from one side of the river to the other. Thus the canyon could not have been the result of some profound fissure or fault that ripped open the earth, only later to be occupied by the Colorado River. Everything in his view was the result of water cutting down into previously unopened earth, even though the side canyons were "dry and bare." This basic observation is the first and foremost lesson that the Grand Canyon reveals to us as and remains such to this day.

Newberry also realized how unique the Grand Canyon landscape was with respect to the earth's increasingly known geography.

> *[These canyons] belong to a vast system of erosion, and are wholly due to the action of water. Probably nowhere in the world has the action of this agent produced results so surprising as regards their magnitude and their peculiar character.*

How strange it seems that Newberry and Ives would respond so differently to the landscape they were charged to explore. Given the fact that they traveled together, slept on the same cold, hard ground, and suffered from the ever-present lack of reliable water sources, it's obvious that one's perspective makes a great deal of difference regarding adversity. It may be a stretch to say that geologists suffer from thirst in a happier state of mind than non-geologists. But given the divergent emotions evoked by seeing the Grand Canyon, it does suggest that an appreciation and understanding of one's surroundings can at least make life more interesting, maybe even more satisfying.

COURTESY OF USGS

John Wesley Powell, 1875

Eleven years would pass before the next geologist laid eyes on the Grand Canyon, and whereas Newberry had the advantage of seeing the Big Cañon from above, John Wesley Powell would benefit by observing it from within. Having spent a few summers naturalizing in the Rocky Mountains of the Colorado Territory, Powell's gaze was ever focused from those lofty heights downstream towards the vast and unexplored heartland of the Colorado Plateau. He was determined to make a thorough investigation of what seemed to be an interesting geography.

In May of 1869, he set out from Green River, Wyoming in four wooden boats with nine men, and began a one-hundred-one-day descent of the Green and Colorado rivers. At the beginning of the trip, Powell had a freshness of spirit that fostered creative scientific thoughts about how the course of the Green River had positioned itself with respect to the Uinta Mountains, located in northeast Utah. By the time he reached the area of the Grand Canyon, he was weary, nearly out of food, at odds with some of his men, and concerned about the treacherous rapids that choked the river. For these reasons, his thoughts on the origin of the Grand Canyon were made only by inference from his more detailed observations on the Green River, as it cuts through the eastern flank of the Uinta Mountains.

An observer looking upstream along the Green River sees a stream meandering through open valleys (top). Turning 180 degrees on that same spot, the terrain is dramatically different at the Gates of Lodore (bottom). Photographs by Wayne Ranney

In his *Exploration of the Colorado River of the West and It's Tributaries*, published in 1875, Powell agreed with Newberry that it was erosion by water, not preexisting fissures, that accounted for the dissection of this canyon country. He took Newberry's idea however, and added to it by suggesting a way in which this erosion had commenced. Powell was struck by the manner in which the Green River ignored and often bypassed low-lying open valleys, only to turn headlong into solid bedrock canyons, such as the Canyon of Lodore on the east flank of the uplifted Uinta Mountains. Common sense dictates that rivers should flow in low valleys and around high mountains, which normally act as barriers to rivers. Yet the Green River did not follow this simple pattern.

To a person studying the physical geography of this country, without a knowledge of its geology, it would seem very strange that the river should cut through the mountains, when, apparently, it might have passed around them to the east, through valleys, for there are such along the north side of the Uintas, extending to the east, where the mountains are degraded to hills, and, passing around these, there are other valleys, extending to the Green, on the south side of the range. Then, why did the river run through the mountains?

The river's relation to the mountains didn't make sense, yet with his geologic training Powell postulated an idea that could explain this rather odd arrangement:

Again, the question returns to us, why did not the stream turn around this great obstruction, rather than pass through it? The answer is that the river had the right of way; in other words, it was running [before] the mountains were formed; not before the rocks of which the mountains are composed were deposited, but before the formations were folded, so as to make a mountain range.

This was a phenomenal leap of insight, not only regarding the immediate problem of the evolution of the Colorado River, but also for the science of geology in general. For perhaps the first time ever, a geologist could look at a confusing set of relationships between a river and the landscape features upon which it was set, and suggest which had formed first. The idea was so new that Powell needed to invent a term to describe it: *antecedence*, meaning that the river's path predates the uplifted features through which it flows.

I have endeavored above to explain the relation of the valleys of the Uinta Mountains to the stratigraphy, or structural geology, of the region, and, further, to state the conclusion reached, that the drainage was established antecedent to the corrugation or displacement of the beds by faulting and folding. I propose to call such valleys . . . antecedent valleys.

To understand antecedence, we can listen to Powell's own explanation of it.

We may say, then, that the river did not cut its way down through the mountains, from a height many thousands of feet above its present site, but . . . it cleared away the obstruction by cutting a cañon, [as] the walls were thus elevated on either side. The river preserved its level, but the mountains were lifted up; as the saw revolves on a fixed pivot, while the log through which it cuts is moved along. The river was the saw which cut the mountains in two.

Antecedence suggests that the river's position is older than the uplift of the mountains. Viewed in this way, the river retained its course even as the land rose up around it, albeit slowly enough so that the river's course was not deflected by the rising mass of rock. Later refinements would challenge the theory of antecedence but Powell provided a clever and thoughtful explanation for how the odd relationship between the Green River and the landscape that enclosed it came to be.

Even so, there was another possibility that could explain what was seen on the Green River and Powell was aware of it. After his two river trips but before the publication of his report in 1875, a geologist named Archibald Marvine mapped the Rocky Mountain region in Colorado. In his report of June 1874, he interpreted a possible sequence of events that gave rise to that landscape. Marvine suggested that the streams there had originated on top of young sedimentary material, which covered and buried an older, mountainous topography. These streams eventually cut through the flat sedimentary cover and chiseled their way down into this buried topography. This is a subtle but significant difference from antecedence. In this case, the river remains the saw but moves down to a buried and stationary mountain mass.

An antecedent river flows across flat terrain (A). As the earth folds, slowly uplifting the terrain, the river cuts a channel to maintain its course (B).

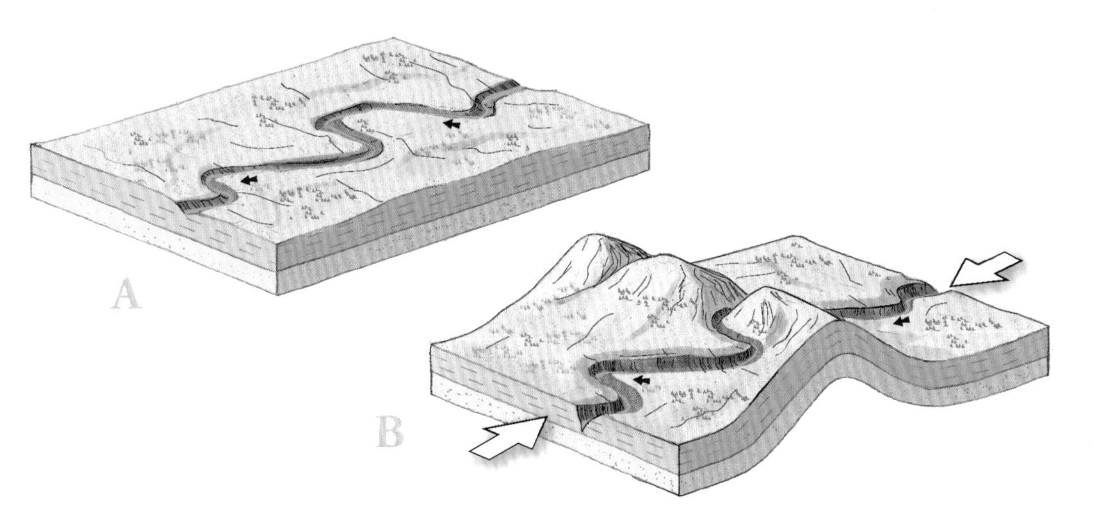

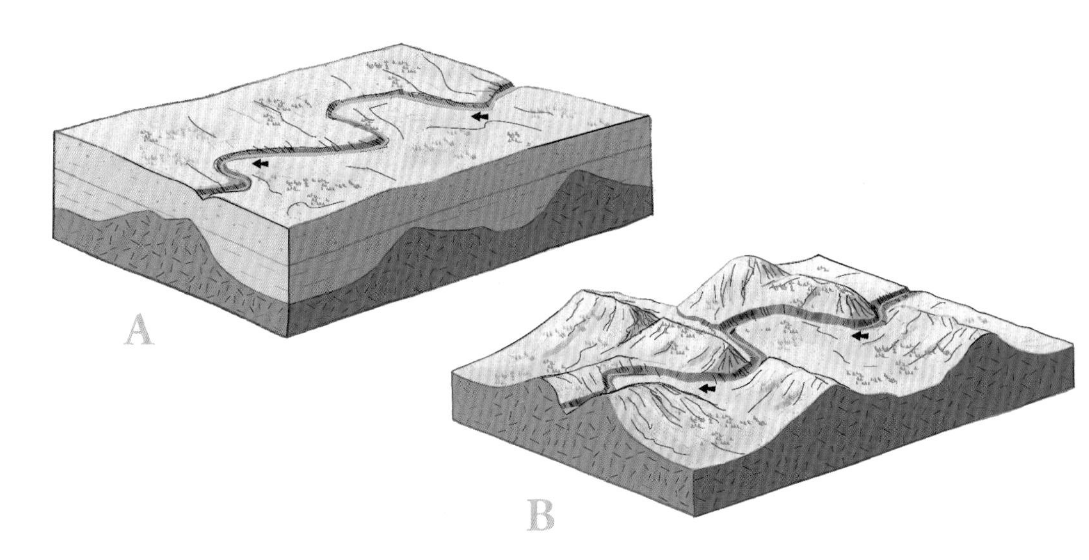

Superposition occurs when a river flowing across a subdued landscape (A) cuts through the soft covering to a preexisting buried topography (B).

Marvine's interpretations seem overshadowed by Powell's rising fame as a successful explorer cum Washington bureaucrat, but Powell himself thought enough of it in his day that he referenced Marvine word for word in his own 1875 report. Powell claimed to have had the same interpretation for the streams in the Rockies from his field work during the summers of 1867 and 1868. He recognized the importance of this idea and proposed a name for it, too: *superimposed valleys*, (today the term is shortened to *superposed*). Superposed valleys are ones which originate on higher, subdued surfaces but work their way progressively down into pre-existing buried landforms.

When Powell wrote his 1875 report however, he insisted that the area of the Green River adjacent to the Uinta Mountains formed by antecedence and not superposition. He seems to have struggled with sorting out the two possibilities, saying that he checked and rechecked the field relations, but ultimately settled firmly in the antecedent camp. One wonders if Marvine took exception to this interpretation, and if so, whether this could have been the first of many differences of opinion regarding the origin of the Colorado River system.

Regarding the Grand Canyon and the Colorado River, Powell actually had little to say about its origin. He did note the apparent lack of association between the many faults in the canyon and the course of the river. With this observation he concurred with Newberry that the canyon was not formed by fissures. However, he made perhaps too great a leap when saying:

All the facts concerning the relation of the water-ways of the region to the mountains, hills, cañons, and cliffs, lead to the inevitable conclusion that the system of drainage was determined antecedent to the faulting, and folding, and erosion.

This was a time when the bigger picture of the West's landscape evolution was completely unknown. Powell didn't have the luxury that we enjoy today of knowing the regional relationships of the river to the geologic stage upon which it was set; his role was simply to establish a framework of generalized ideas. More importantly, he couldn't have given much thought to the origin of the Grand Canyon simply because he was in a race for survival against it. He took what he more carefully observed upstream on the Green River and tried to make it fit with the little that he did observe in the Grand Canyon. His pioneering explorations served to mentor and inspire two successive generations of southwestern geologists.

COURTESY OF USGS

Clarence Dutton, 1882

By the early 1880s, at least two geologists had observed the Grand Canyon—one from its south side, the other from along the river. How appropriate then that the next geologist to ponder its development would do so from the North Rim.

Clarence Dutton, a Yale-educated army officer, immediately recognized the global significance of the landscape laid out before him at places like Point Sublime and Toroweap. He approached the Grand Canyon after traveling through the High Plateaus in southern Utah and was able to make the important evolutionary connection between the two. He recognized that strata composing the Grand Staircase (a name he invented), once covered the Grand Canyon region, only to be stripped away in what he called the Great Denudation. He theorized a later period of canyon cutting, which he

Artist William Henry Holmes accompanied Clarence Dutton on his survey of the Grand Staircase and Grand Canyon's Toroweap area and painted the first staggering images of this unique landscape. The people and horses near the waterpocket provide scale. *The Grand Cañon at the Foot of the Toroweap–Looking East* by William Henry Holmes, 1882

termed the Great Erosion. Dutton therefore, was the first geologist to differentiate between two cycles of erosion: one that created the Grand Staircase by the lateral stripping of strata and one that created the Grand Canyon through vertical dissection. These two very different periods of erosion led to the landscape seen today.

Clarence Dutton wrote what is perhaps the most readable, maybe even lyrical, scientific report ever written—*The Tertiary History of the Grand Cañon District*. Published in 1882 as a monograph of the United States Geological Survey, this report summarized his observations in which he concurred with Powell for an antecedent origin of the river.

The river is older than the structural features of the country. Since it began to run, mountains and plateaus have risen across its track and those of its tributaries. . . . As these irregularities rose up, the streams turned neither to the right nor to the left but cut their way through in the same old places.

However, he added something original of his own to Powell's version which spoke to the manner in which the Colorado was placed in its exact course.

What then determined the present distribution of the drainage? The answer is that they were determined by the configuration of the old Eocene Lake bottom at the time it was drained.

This is a notable addition to the growing litany of ideas regarding the possible evolution of the Colorado River. Here, Dutton suggests a superposed origin for the river's precise location upon the landscape. He states that when the Eocene (a subdivision of the Tertiary Period, lasting from 58 to 37 million years ago) lakes dried up, the river flowed across the low spots of its bed. He returns to antecedence later however, by stating again that the folds and faults became active after the course of the river had already been determined by the shape of the lake bottom.

Dutton was one of the greatest minds to ever ponder the origin of the Grand Canyon landscape but it was beyond him to formally challenge the prevailing view of his mentor regarding the theory of antecedence. However, Dutton's greatest gift to us may be his appreciation of the Colorado Plateau landscape, for he was perhaps the first person to view it as intrinsically beautiful or even nurturing. His writing signals a collective change in the view all subsequent persons would have regarding this exotic landscape.

The lover of nature whose perceptions had been trained in the Alps . . . or . . . the Appalachians . . . would enter this strange region with a shock, and dwell there for a time with a sense of oppression, and perhaps with horror. Whatsoever things he had learned to regard as beautiful and noble he would seldom or never see, and whatsoever he might see would appear to him as anything but beautiful and noble. . . . The colors would be the very ones he had learned to shun as tawdry and bizarre. . . .

But time would make a gradual change. Some day he would become conscious that outlines which at first seemed harsh and trivial have grace and meaning; that forms which seemed grotesque are full of dignity . . . that colors which had been esteemed unrefined, immodest, and glaring, are as expressive, tender, changeful, and capacious of effects as any others. Great innovations, whether in art or literature, in science or in nature, seldom take the world by storm. They must be understood before they can be estimated, and must be cultivated before they can be understood.

COURTESY OF USGS

Charles D. Walcott, 1890

Charles Walcott is most famous for his work in the unusual fossils of the Burgess Shale in the Canadian Rockies. However, in 1882 his field party constructed the Nankoweap Trail in Grand Canyon, which gained him access to the faults and folds in eastern Grand Canyon. The Butte Fault, which he named, marks the eastern edge of the Kaibab upwarp and he determined that this fault and the associated East Kaibab monocline, were actively uplifting the Kaibab Plateau before the overlying Mesozoic rocks were stripped away to the northeast. This hinted that the fold could be much older than previously thought. He writes in his report:

> ***It is difficult to understand how the cañon could have existed even to a limited depth, in its present position, at the time of the elevation of the Kaibab Plateau. . . . If followed out in all its bearings, [this] would probably necessitate some change in the now accepted views concerning the manner of erosion . . . of the Grand Cañon.***

Charles D. Walcott mapped and named the Butte Fault (center), which marks the eastern boundary of the Kaibab upwarp. Walcott recognized that the fault was older than the river. Photograph by Larry Lindahl

Walcott never overtly challenged Powell's theory of antecedence but his words suggest a discomfort with the idea that would become a firestorm of doubt in the not-too-distant future. In his report he noted how the Colorado River in Marble Canyon exactly parallels the trace of the Butte Fault for ten miles from Nankoweap Creek to the Little Colorado. This showed unequivocally that the river must have become positioned after that structure was formed. In this way, he became the first geologist to significantly challenge the prevailing idea that the Colorado River was older than the faults and folds that cross its path.

S. F. Emmons, 1897

If C. D. Walcott delivered the first quiet shot to Powell's theory of antecedence for the Colorado River, then S. F. Emmons, of the U.S. Geological Survey, fired the cannons. He published a short article in the journal *Science* based on work he completed in the area of the Uinta Mountains and the Green River in 1872. In it he vigorously sought an explanation from Major Powell on how the Green River could be older than the Uinta Uplift, when eight thousand feet of Tertiary lake sediment was derived from the uplifted mass, and the Green River later cut through those same sediments.

What then became of the river while these eight thousand feet of Tertiary sediments were being deposited? It could hardly have continued its course at the bottom of the . . . lakes. . . . If it ceased to flow during this time its bed must have been filled with sediments . . . and when the lakes finally drained, it is hardly conceivable that in re-determining its course . . . it should have attacked the flanks of the Uinta range . . . at exactly the same point it had entered before.

In this passage, Emmons offered a convincing idea about the age of the Uinta Mountains relative to the Green River. He pointed out that the river cut through lake deposits that were derived from the adjacent uplift. This suggested to him (and many that followed) that these deposits were younger than the uplift and, since the river sliced through these sediments, the river was younger than the sediments. Emmons writes that he tried to engage Major Powell in an explanation for his line of reasoning but ". . . he says it is so long ago he no longer remembers the course of reasoning he followed at the time." Emmons concludes, "Whatever may be the outcome of such an examination, it would seem proper that the antecedence origin of this river should be held in abeyance until some positive evidence of it can be furnished."

The nineteenth century closes with the first attack on the theory of antecedence for the origin of the Colorado River, yet no alternative is poised to take its place. A basic framework of ideas has been embraced but the increasing recognition of the relationships between the river and the landscape that held it, required a change in thinking that would shake the very foundations of antecedent thought.

Early Twentieth Century

The twentieth century dawned with geologists undertaking field excursions to the Grand Canyon on horse and buggy, and closed with astronauts taking photos of the canyon from space as they circled the earth from above. One might surmise that this degree of change in technology would spill over to a similar degree in our understanding of the canyon's origin. In many cases, it does not. Many of

Vermilion Cliffs, seen on the horizon from this North Rim vantage, are part of the Grand Staircase. Photograph by Chuck Lawsen

the same dilemmas that confronted early twentieth-century geologists still haunt us. How old is the river? How did it carve the canyon? What is the timing of plateau uplift? Geologists who came to the river's edge downstream from Grand Canyon during the construction of Hoover Dam became interested in these questions. They looked at the relationship between the river and its associated deposits before the deposits were inundated forever by the rising waters of Lake Mead. Ironically, it was the obstruction of the free-flowing Colorado that led to important insights into its origin and evolution.

William Morris Davis, 1901

In June of 1900, William Morris Davis, known as the father of geomorphology (the study of landforms), completed a twenty-three-day overland trip from Flagstaff, Arizona to Milford, Utah. He traveled by horse and wagon, averaging twenty-five miles a day in this manner. His route took him north to the Echo Cliffs and Lees Ferry, then along the Vermilion Cliffs and over the Kaibab Plateau to Fredonia, where he proceeded down the Toroweap Valley to a stunning view of the Colorado River. Davis was an astute observer of landscapes and took a keen interest in how the Grand Canyon may have formed.

Davis was thankful for the foundation that his nineteenth-century colleagues had bequeathed to him but he also suggested ways that their findings could be refined. He quantified the idea first proposed by Dutton that two cycles of erosion were responsible for the present arrangement of deep canyons set within broad plateaus. The first cycle he termed the "plateau cycle" which correlates with Dutton's Great Denudation and the lateral stripping of the Grand Staircase to the north during the early Tertiary. His second cycle, called the "canyon cycle," was correlative with Dutton's Great Erosion and was of more recent origin, being the time when the deep incision of the canyons took place. Broadly, he argued for a humid climate during the "plateau cycle" and an arid climate for the "canyon cycle." The climatic relationship to erosion style is an idea that is still valid today.

Regarding the theories of Powell and Dutton, Davis was not comfortable with a purely antecedent origin for the river and argued that it might be consequent to (positioned in response to the ancient landscape features) the presence of the faults, folds, and topography that previously occupied the area. He writes:

It is not my intention to discount such [theories] . . . but only to emphasize the opinion that the facts now on record, combined with such knowledge of the region as our party was able to gather . . . warrant the consideration of at least one hypothesis alternative to the theory of antecedence, as an explanation for the origin of the drainage lines in the Grand Canyon district. I do not on the one hand consider the antecedent origin of the Colorado disproved, but, on the other hand, such an origin does not seem compulsory. The chief objection to the theory of antecedence is not that rivers cannot saw their way through rising mountains . . . but rather that this theory makes a single stride from the beginning to the end of a long and complicated series of movements and erosions, overlooking all the opportunities for drainage modifications on the way.

Davis's words reflect an endearing diplomacy towards Powell's theories, while at the same time it challenges them. According to Davis, drainage in the Grand Canyon area was initially directed northeastward down the steps of the east-dipping monoclines and towards the Grand Staircase. Later, as the Basin and Range Disturbance dropped fault blocks down to the west, this drainage was reversed by both the lowering of the Basin and Range and reversed tilting of the entire plateau landscape. The deep dissection of the canyon occurred after Basin and Range faulting and the reversal of the drainage. These were bold concepts that reinvigorated thinking on the origin of both the river and the canyon and many of them may still ring with truth today.

Davis also offered cogent observations into the possible origin of the Esplanade, saying that it was not the result of the river meandering freely across that surface for an extended period of time before the deep dissection of the inner canyon. Instead, he pointed out that everywhere the Esplanade is found within Grand Canyon, it is always equidistant from either the north or south rim. He postulates that it was the inherent softness of the Hermit Formation upon exposure to erosion that caused the upper cliffs on both sides of the canyon to retreat at an equal rate away from the incising river. His interpretation is likely correct.

He also noticed with interest that almost all tributary streams in the Grand Canyon enter the Colorado River at grade (at the same level), even though many of them are dry most of the year. He suggested that this could be the result of erosion rates having been greater at some earlier time; otherwise how could a stream that is dry most of the time erode as deeply as the Colorado River. His recognition of the way these tributaries enter the main stream is a phenomenal insight. Although we can now explain the lack of "hanging valley" tributaries along the Colorado River for other reasons, this observation of these relationships allows us to ponder how streams, even dry ones, can deepen their canyons.

All of these observations prompted Davis to state that even though the canyon was extremely deep and rugged, the Colorado behaved essentially as a mature system within it. This is verified by noting the relatively low gradient that the Colorado River has today within Grand Canyon, averaging about eight feet per mile. It appears that the river is not excavating actively into the bedrock to any great degree in the modern setting. All of these insights were fresh and original and stimulated others to look critically at the evidence for how long the river might have been at work in carving the Grand Canyon. Davis offered a synthesis of what he observed on this trip and a second field trip conducted in 1902:

The most emphatic lesson that the canyon teaches, is that it is not a very old feature of the earth's surface, but a very modern one; that it does not mark the accomplishment of a great task of earth sculpture, but only the beginning of such a task; and that in spite of its great dimension, it is properly described as a young valley.

In the first decade of the twentieth century other geologists contributed ideas dealing with the origin of the Colorado River. Willis Lee (1906) proposed that the river may have once flowed through the Sacramento and Detrital Valleys, located between Kingman and Las Vegas. He used the presence of gravels in those valleys as evidence. Harold Robinson (1909) documented the existence of a broad erosional surface that was formed on and adjacent to the Colorado Plateau before the canyon cycle of erosion. According to him, a part of this old surface is preserved under the lava flow on Red Butte, south of Grand Canyon. Douglas Wilson Johnson, on a wagon trip in 1906 from Prescott, Arizona to Salt Lake City, Utah, verified Walcott's and Davis's reports that the faults and folds of the region seem to be older than the river. As more eyes looked upon the canyon landscape, there began to emerge the idea that the Colorado River might have attained its present position later than previously thought.

COURTESY OF USGS

Eliot Blackwelder, 1934

Most geologists up to this point had argued for a Colorado River that was old—at least early Tertiary in age (more than 37 million years old). Eliot Blackwelder would offer a challenge to this generally accepted view. He sought to dismiss Lee's postulation that the river once flowed south through the Sacramento and Detrital Valleys, by showing that the gravels there were only of local derivation and not deposited by the river. He looked at the entire river system from its headwaters to its mouth and recognized its curious setting through open valleys that alternate between narrow, confined canyons. With these observations, he sought to advance the idea that the Colorado might actually be a young river.

Blackwelder noted the presence of closed basins immediately adjacent to the river south of Las Vegas that were not part of the river system. He argued that if the river was old, why had it not captured these basins by extending its tributaries up into them? He looked at other southwestern streams such as the Owens, Amaragosa, and Sevier Rivers and noticed how they had developed their courses by filling basins along their path, which overflowed successively into the next basin, until a string of lakes was present—all connected by steep-gradient rivers that were confined in narrow canyons. Blackwelder envisioned a pre-Colorado River environment that was essentially arid and replete with intermontane basins. With the uplift of the Rocky Mountains, a catchment for moisture was created that then flowed south into these interior basins.

The present existence of the Colorado River is due solely to the fact that the Rocky Mountains in Colorado, Wyoming, and Utah are sufficiently high to condense moisture. . . . It is reasonable to infer that, as the region bulged upward, the local streams on the higher and more northerly mountains extended themselves [southward], forming lakes in the nearest desert basins. As this influx exceeded evaporation . . . the lakes rose until they overflowed the lowest points of their rims and spilled into adjacent basins. In time, enough excess outflow may have developed to fill a series of basins all the way to the Gulf of California, thus forming a chain of lakes strung upon a river.

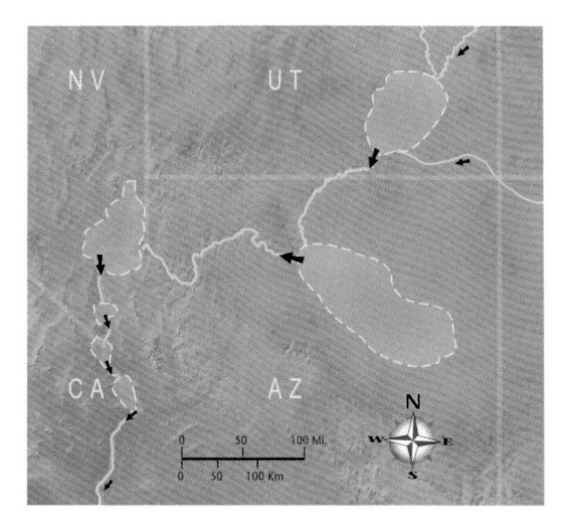

Blackwelder envisioned a string of lakes along the course of the Colorado River from the Rocky Mountains to the sea.

Blackwelder knew that such a postulation was highly speculative but used what he saw in the present-day landscape to formulate a viable story for a young Colorado River. While it may be a stretch to consider this relatively simple hypothesis for a system as large and complex as the Colorado, he did give a plausible alternative to the idea that the Colorado River must be ancient. Ironically, as we shall see, significant similarities for his idea of a "fill and spill" evolution of the Colorado River both above and below the Grand Canyon are enjoying a renaissance. Blackwelder's closing argument is perhaps the most cogent summary given to this point:

The foregoing sketch of the origin and history of the Colorado River is frankly theoretical. Science advances not only by the discovery of facts but also by the proposal and consideration of hypotheses, provided always that they are not disguised as facts. This view will not meet with general acceptance. There are doubtless many facts unknown to me that will be brought forward in opposition. Perhaps their impact will prove fatal to the hypothesis. In any event, the situation will be more wholesome, now that we have two notably different explanations, than it was when it was assumed by all that the river had existed continuously since middle or early Tertiary time. It seems to me that the new hypothesis is harmonious with most of the important facts now known about the geology and history not only of the Colorado River but of the Western States in general.

Eliot Blackwelder remains one of the giants in the evolution of ideas regarding the origin of Grand Canyon. He challenged a prevailing view that most geologists had accepted up to this point. And as we shall see, an increasing number of modern geologists believe that overflow of ponded sections athwart the river's course may be how the integration of the river system was accomplished.

Donald Babenroth and Arthur Strahler, 1945

Up to this point, few had specifically addressed the problem of why the Colorado River abruptly turns ninety degrees into the barrier-like Kaibab upwarp in eastern Grand Canyon. Geologists had long recognized this confusing relationship between the river and the upwarp but no one had tackled the problem head on. That would change in 1945 when Donald Babenroth and Arthur Strahler published a paper that stated:

The feature requiring special attention is that the river passes from the relatively low Marble Platform area westward through the Kaibab arch which resembles a giant anticlinal barrier in the path of the river.

This idea focuses on the structure of the East Kaibab monocline and how it may relate to the possible origin of the Colorado River. First, the authors reviewed the ideas of Powell, Dutton, Walcott, and Davis. They discarded Davis' idea of a regional reversal of tilting of the plateau and the river channel, stating that the amount of reversal necessary was too much given the evidence at hand. This was an important rejection which allowed subsequent thought to focus more clearly on other mechanisms for how drainage reversal of the river was accomplished. Babenroth and Strahler inferred that the river, since its inception, may have always flowed across the Kaibab upwarp, favoring a subsequent origin for the river's position across the upwarp.

In general plan, the Colorado River makes a great bend [westward] around the pitching nose of the Kaibab arch. This suggests that, as the Mesozoic strata were being stripped from the region, the river may have occupied a subsequent lowland belt of weak Moenkopi and Chinle shales between the plunging nose of the Kaibab limestone arch and the encircling north-facing cliffs of Jurassic sandstones. . . . The

subsequent valley would have been several miles wide and would have coincided approximately with the present Grand Canyon.

Babenroth and Strahler explained how the Colorado River could have been positioned in its present course on the south flank of the Kaibab upwarp. They envisioned that the river was confined in a low valley, formed between the south facing slope of the Kaibab upwarp (essentially the surface of the North Rim today) and Mesozoic rocks that were being stripped off this surface towards the south (essentially where the South Rim is located, with Red Butte and Cedar Mountain as the only visible remnants of this once more extensive cliff.) Their theory required that at the time of positioning across the Kaibab upwarp, the Marble Platform be buried by a thick succession of Mesozoic rocks, such that the upwarp was not differentiated topographically from the lands that surrounded it. How else could the river flow downhill towards the west?

Babenroth and Strahler offered the first explanation for how the river may have been placed across the Kaibab upwarp. Later, a well-known geologist would pick up on this idea and popularize it with the moniker known as the "racetrack" theory.

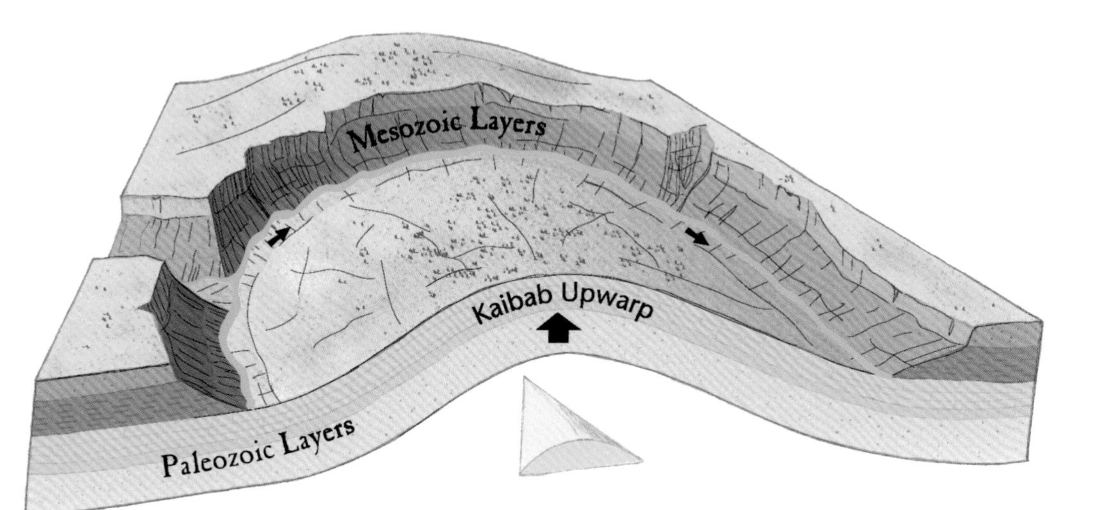

Babenroth and Strahler suggested how the Colorado River may have been positioned. They envisioned a river flowing through a subsequent valley between a receding cliff of Mesozoic rock and the arch of the Kaibab upwarp.

COURTESY OF USGS

Chester Longwell, 1946

Chester Longwell was a professor of geology at Yale University in the 1930s and studied deposits in the Grand Canyon area that would become flooded by Boulder (later Hoover) Dam, beneath today's Lake Mead. Longwell was fascinated by the emerging dichotomy regarding the differing opinions on the age of the Colorado River.

One of the major unsolved problems of the region is the date of origin of the river itself. . . . Geologists who have no direct acquaintance with the region will be at a loss to understand so wide a divergence in interpretation.

How true his words ring even today! In his paper, "How Old Is the Colorado River," Longwell summarized the two extreme views: those of Eliot Blackwelder who believed that the river developed from a string of over-topped lakes in only the last 2 million years, and those of other geologists who used evidence from Colorado and Utah for a river perhaps 40 to 60 million years old. Longwell looked at the deposits of the Muddy Creek Formation just west of the Grand Wash Cliffs, and concurred with the earlier findings of Chester Stock (1921) that they were not derived from the Colorado River. The deposits are as young as 6 million years and Longwell's conclusion was that the river, and thus the Grand Canyon, could be no older than this. Contrasted with the views of those favoring an older river, the evidence from the Muddy Creek Formation was increasingly referred to as "the Muddy Creek problem." Longwell elaborated on the importance of the evidence:

There is no possibility that the river was in its present position west of the plateau during Muddy Creek time. The suggestion occurs that a stream, either permanent or intermittent, may have developed on the site of the present Grand Canyon, and debouched into closed basins west of the plateau. However, if such a stream had any considerable length, it should have contributed rounded pebbles representing the varied lithology east of the Grand Wash Cliffs. No such stream-worn pebbles have been found in the basin deposits. . . . Deposits made by the through-going Colorado are unmistakable. They contain, as a characteristic ingredient, gravel made up of rounded pebbles and cobbles, representing a large number of bedrock types found in the plateau.

The western end of the Grand Canyon is now inundated by Lake Mead. In the background are the Grand Wash Cliffs; partially eroded deposits of the Muddy Creek Formation fill the foreground. Photograph by Wayne Ranney

This critical piece of evidence has served since the mid-twentieth century as the pivot point around which all ideas concerning the age of the Colorado River have revolved. As we shall see, later geologists would be required to devise clever and intricate ways for the Colorado River to flow away, around, or underground from the Muddy Creek problem. Longwell's observations remain the funnel through which other theories must flow.

Longwell also noted a very important piece of information regarding the relationship of the Colorado River's development to the larger development of the western landscape.

Attempts to project the present drainage system as far back as early Tertiary encounter possible difficulty in some significant facts of the regional history. Intensive orogeny [mountain building], with development of great thrusts and folds . . . affected a wide belt directly west of the plateau near the close of the Cretaceous Period [144-65 million years ago]. Presumably, therefore, a mountain zone extended across the course of the present Colorado River in early [Tertiary] time, with . . . topography opposed to any westward drainage from the lower plateau area.

What Longwell refers to is the existence of the Mogollon Highlands, a range of mountains that existed south and west of the Colorado Plateau until the Basin and Range Disturbance destroyed them beginning 16 million years ago. Longwell asks, What is the likelihood that an early incarnation of the Colorado River would have flowed towards these uplifted mountains? The answer is easy, it could not. The crucial point is that any precursor to the Colorado River could not have been flowing west before the Basin and Range developed, beginning 16 million years ago. More probable is that drainage on the southern Colorado Plateau was directed to the northeast and possibly ponded there in enclosed basins. Longwell astutely pointed out that many geologists seemed to have ignored this salient fact and he speculated that:

> *In outlining the foregoing hypothesis, it has been assumed that the plateau has had exterior drainage continuously though the [Tertiary period]. However . . . the region probably was unable to support a through-flowing stream like the Colorado for a considerable period after the onset of aridity in the western interior. . . . During such an interval the drainage of the plateau area would have been accomplished by intermittent streams ending in a number of separate closed depressions, as in the Great Basin at present. . . . When the [Rocky Mountains] attained such altitude that increased precipitation, [they] supplied a surplus of runoff into the plateau, the configuration of the surface may have been such as to guide the overflow along a new consequent course to the west.*

Longwell appears to be lining up behind the idea presented by Eliot Blackwelder before him. He agreed that the plateau landscape was perhaps dominated by closed, interior basins before the modern Colorado River was born. Their historical vision is one that seems to be enjoying a welcome revitalization in the current discussion of the river's evolution.

COURTESY OF USGS

Herbert Gregory, 1947

Since the beginning of the twentieth century, many workers concerned with the origin of the Colorado River had seen evidence for the existence of plateau-wide erosion surfaces that presumably formed during the period of lateral stripping and before the carving of deep canyons. Davis, Robinson, and Herbert Gregory were among these workers. Gregory however, who accompanied Davis on his 1900 expedition and was an erosion-surface advocate himself (1917), had changed his mind by 1947 after many field seasons scampering across the plateau. He believed that the perceived surfaces were either of local importance only (pediments) or were formed only as the result of soft strata being removed from flat, underlying resistant beds, thus forming surfaces that only appeared regional in nature. The distinction between a truly eroded surface and one formed because of stratigraphic resistance remains largely unresolved today.

Gregory's observations relating to the history of the Colorado River suggest that it was the uplift of the High Plateaus in Utah that provided the elevation and location necessary for precipitation to establish the modern river.

Guided by topography, rains and melting snows doubtless spread a net of short . . . streams over the plateau tops. . . . It may be conjectured that the south-flowing waters from the uplifted, warped, and fractured highland found a resting place on the structural lowlands in the approximate position of the present Grand Canyon and the . . . Little Colorado Valley; that their most easily developed outlet to the sea was westward; and that along this course the new-born stream, locally expanded into lakes, descended from one to another plateau block over high cliffs, and thus developed energy rarely acquired by running water.

Gregory's statement that the waters found a resting place in the position of the Grand Canyon and the Little Colorado Valley is an interesting one. One assumes that he appealed to "a resting place" on either side of the Kaibab upwarp, for he continues:

It seems probable that the original Colorado was a relatively short stream of moderate volume fed from the north by the ancestral Virgin, Kanab, and Paria rivers, and from the south by the Little Colorado, which gathered water from the large region between Navajo Mountain and the Mogollons.

Gregory believed that the "young" Colorado was able to capture drainage area from other nearby rivers because of its steeper gradient and its shorter course to the sea towards the Gulf of California. In this way, according to him, it became the integrated river we find today. He may have been the first geologist to overtly propose the idea of stream capture by headward erosion in the debate about the Colorado's origin.

The first half of the twentieth century ends with geologists increasingly arguing for a young Colorado River. These ideas contrast with the earliest views that the river was older. The next fifty years would attempt to resolve these seemingly conflicting views.

Legends and Non-Scientific Ideas

Long before scientists thought about the origin of the Grand Canyon, native peoples were sharing legends about how the canyon was created. Indian legends are not scientific theories but curiously, as we shall see, they do parallel some modern ideas that are. If any of us had been alive just over one hundred fifty years ago and belonged to this heritage, these legends would absorb our thoughts as we gathered around the campfire on a cold winter night and listened to our elders recall the creation days.

Five modern tribes live in or adjacent to the Grand Canyon—the Havasupai, Hualapai, Southern Paiute, Navajo, and Hopi—and several other tribes maintain close ties to the canyon. Each group has its own creation story which has been handed down and protected through the centuries.

Other native peoples preceded these tribes and archeological evidence attests to their former presence, but nothing remains of what their thoughts might have been regarding the origin of the Grand Canyon. The earliest evidence of human occupation in the Grand Canyon dates back four thousand years, but what these split-twig figurine makers thought about the canyon will probably never be known to us. Perhaps even the mammoth hunters were here almost twelve thousand years ago, but the evidence is scant. Imagine the first human being ever to lay eyes on the Grand Canyon some twelve thousand years ago!

Ancestral Puebloan petroglyphs in a dry wash along the Colorado River near Tanner Rapid. Photograph by Larry Lindahl

Native American creation stories relating to the Grand Canyon typically invoke catastrophic floods, a theme common for many agricultural groups who live along rivers throughout the world.

"One day a mischievous god, named Ho-ko-ma-ta, started a rainstorm greater than a thousand Hack-a-tai-as (Colorado Rivers). The benign god, Toc-ho-pa, wanted to save his daughter, Pu-keh-eh, from the ensuing flood and put her in a hollow piñon log. As the floodwaters rushed to the sea, they created Chic-a-mi-mi (Grand Canyon) and Pu-keh-eh was able to crawl out of her log safely and become the progenitor of all human beings."

This creation story was recorded by an anthropologist studying the Havasupai. It is a beautiful story with a number of parallels to contemporary scientific theories about how Grand Canyon was formed. The story astutely associates the cutting of Grand Canyon with the Colorado River. This is the basic tenet of the geologic story and if we were required to summarize the entire story of the Grand Canyon into one short sentence it would be The Colorado River carved the Grand Canyon. The legend also describes the whole process as happening during a single flood and on a catastrophic time scale. Some modern geologists invoke catastrophic events like this for the creation of the river and the canyon we see today. Other tribes have their own creation stories which also parallel geologic ideas in some ways.

In 2003 a new book was published documenting a religious interpretation for how the canyon was formed. Its premise is that all of Grand Canyon's flat-lying strata, along with those still found at Zion and Bryce Canyon national parks, were deposited during Noah's flood. Recession of this water completely removed the Zion and Bryce Canyon rocks from the Grand Canyon area and carved the great gorge quite catastrophically. According to this story, all of these events—the deposition of 15,000 feet of sediment, complete removal of the upper half of it and the cutting of the Grand Canyon—happened during a single flood in a *one-year* time period. Geologists argue for a much longer timespan for the deposition of the strata and a completely separate set of events at a much later time to carve the canyon. The idea that all of this geology could have occurred in one year is inconceivable to most scientists or anyone who has critically looked at the evidence for rates of change in landscape development.

More troublesome to traditional scientists however, is the manner in which this "creation research" was conducted. True science does not proceed by drawing conclusions first (there was a flood that covered the whole earth), then identifying those small pieces of the evidence that make a conclusion seem sound (the Grand Canyon must be a result of this flood). A scientist looks at all the possible ways a landscape could have evolved and then tests his or her hypothesis against evidence to determine whether it is sound. In this way, some theories will survive the test of rigorous examination while others will not. "Creation research" is based heavily on religious faith and the belief that the *Bible* is the true word of God, which cannot be challenged. It also assumes the *Bible* can be used as a text to know Earth history, an idea that is troublesome even to many people of faith. Contrary to what creationists say, this faith-based approach is not science. As with all ideas, this material should be approached with an open mind, studied carefully, and critically considered.

Late Twentieth Century

During the second half of the twentieth century, field studies related to the origin of the Grand Canyon increased tremendously. The search for uranium, prompted by the beginning of the Cold War, initiated numerous field studies on the plateau and some geologists became interested in the origin of this captivating landscape. One of the watershed events during this time was a symposium held at the Museum of Northern Arizona that focused specifically on the origin of the Colorado River in Arizona.

COURTESY OF USGS

Charles Hunt, 1956

One of the great names in the field of Colorado River studies is Charles (Charlie) Hunt of the U. S. Geological Survey. His classic paper, "Cenozoic Geology of the Colorado Plateau," was a synthesis of information known about the river and its landscape, from the Uinta Mountains to the Mojave Desert. He postulated the development of the Colorado River by constructing ten paleogeographic maps that showed how the river may have altered its course through time. Hunt offered some intriguing conjectures on the origin of the Grand Canyon as well.

Like many of his predecessors, Hunt recognized the overwhelming evidence for initial northeast flow of drainage across the southern Colorado Plateau. His maps indicate at least three different freshwater lakes where these streams may have terminated. He reminds us that a long period of time, in which no deposits are preserved on the plateau, makes all interpretations necessarily conjectural. Although he accepts that short local drainages may have appeared at the west edge of the plateau during middle Tertiary time, (about 30 million years ago), water was probably still ponded on the central plateau. About this time he suggested that the plateau was uplifted en masse and this is probably when superposed tributaries like the San Juan were entrenched in their canyons. He believed that the rivers continually adjusted their courses as subsurface intrusions of magma formed the laccolithic mountains of the

Plateau—the Henry, Abajo, La Sal, and Sleeping Ute mountains.

According to him, as the plateau became higher than the Basin and Range, Hunt insisted that drainage had to develop off of the plateau edge. But where? He accepted that it couldn't have been at the Grand Wash Cliffs because of the "Muddy Creek problem" defined first by Longwell. So Hunt hinted that the Colorado River in Grand Canyon may have flowed south through Peach Springs Wash. He was quite cautious, however, about this idea.

Hunt was left with two possibilities for the origin of the river in Grand Canyon—superposition or stream capture—and he didn't like either of them. Superposition demanded that lake sediments be present as high as the top of the Kaibab Plateau, presently at over nine thousand feet, and then rise even higher upstream along the river so that it could flow to the southwest. This was untenable to him. Stream capture was also problematic because:

It would indeed have been a unique and precocious gully that cut headward more than 100 miles across the Grand Canyon section to capture streams east of the Kaibab upwarp.

Ultimately, Charlie Hunt had to invent a process and a name for it to get around the dilemma of the "Muddy Creek problem." He called this process *anteposition.* It incorporated aspects of both antecedence and superposition but has never really gained favor within the geologic community. If nothing else it shows to what extremes geologists will go to explain an enigmatic river whose history is essentially destroyed.

Anteposition suggests that the current path of the river through the Grand Canyon was initially established before the Muddy Creek Formation accumulated. Uplift of the plateau then tilted the river's channel towards the east, disrupting and halting its flow into the Muddy Creek basin. This is how Hunt explained the lack of Colorado River sediment within the Muddy Creek Formation. He postulated that the river then became ponded north and east of Grand Canyon, where lake sediments like the Bidahochi Formation accumulated. As this lake filled, it overflowed to the west and clear water spilled from it, depositing the Hualapai Limestone. Eventually, the Colorado River re-established its old course by superposition on the lake sediments to the east of Grand Canyon and it made its way through the Grand Wash Cliffs and began dissecting the Hualapai Limestone.

Hunt's ideas were much debated because they were presented at a time when few other original ideas were forthcoming. In the end, however, it seems that he had but few converts.

Symposium on the Cenozoic Geology of the Colorado Plateau in Arizona, 1967

No treatment of the origin of the Grand Canyon is complete without a discussion of the innovative ideas developed at the Symposium on Cenozoic Geology of the Colorado Plateau in Arizona. The symposium was held at the Museum of Northern Arizona (MNA) in Flagstaff, Arizona in August, 1964, where for the first time geologists gathered in one location for the sole purpose of discussing how the Colorado River may have evolved and perhaps more importantly, what problems remained to be resolved. The symposium was chaired by Dr. Edwin "Eddie" McKee, a prominent Grand Canyon geologist who through his long, distinguished career worked alternatively for Grand Canyon National Park, MNA, the U. S. Geological Survey, and the University of Arizona. Twenty eminent geologists attended the ten-day conference to share with one another their theories, conjectures, and ideas for further research. Sixteen geographic areas in northern and central Arizona were reported on and resulted in two significant milestones: the development of a timeline that outlined a plausible sequence of events for the region, and an original and provocative theory regarding how the Colorado River and the Grand Canyon may have formed from the integration of two separate and distinct river systems. These ideas were published in 1967 as MNA Bulletin #44, "Evolution of the Colorado River in Arizona."

COURTESY OF MNA

Eddie McKee

MNA Bulletin #44, 1967. Courtesy of the Museum of Northern Arizona

In this seminal report, five stages in the development of the Grand Canyon landscape were proposed:

1) Original northeast drainage on a subdued but sloped surface towards a retreating seaway
2) Slight modification of this drainage pattern around the rising upwarps of the Colorado Plateau and flowing into freshwater lakes in Utah, Wyoming, and Colorado
3) Renewed uplift of the plateau and the development of two separate drainage systems; each on either side of the Kaibab Plateau
4) Existence of ponded drainage (Bidahochi and Muddy Creek basins) east and west of the Grand Canyon
5) Development of the modern drainage by the integration of the two river systems, facilitated by continued uplift, headward erosion, and stream capture.

STAGE 1 refers to the "blank canvas" that was the northern Arizona landscape just after the sea retreated for the last time some 80 million years ago. Left in the sea's wake was a surface gently sloping down towards the northeast. Thus, initial drainage in the area was set in a direction opposite to that of the modern Colorado River. This may appear to be a puzzling beginning for such an important river but ironically, this scenario is one of the few aspects of the Colorado River's story that is not controversial among geologists!

STAGE 2 represents only a minor modification on this landscape with the differential uplift of structures like the Kaibab upwarp and the Rocky Mountains. During this stage drainage continued to the northeast, flowing towards freshwater lakes that formed in the basins between the various upwarps. Lake deposits from this stage remain only as far south as the Bryce Canyon area in southern Utah.

STAGE 3 The development of two separate drainages in which the Kaibab Plateau acted as a drainage divide. The western system was called the Hualapai drainage, since its location was near the plateau of the same name. According to the authors this system drained west into the emerging Muddy Creek basin. The other drainage

A diagram from MNA Bulletin #44 shows (A) that the Kaibab upwarp separated the ancestral upper Colorado River from the Hualapai drainage; (B) how headward erosion of the Hualapai drainage captured the Colorado River; and (C) subsequent deepening by the integrated stream.

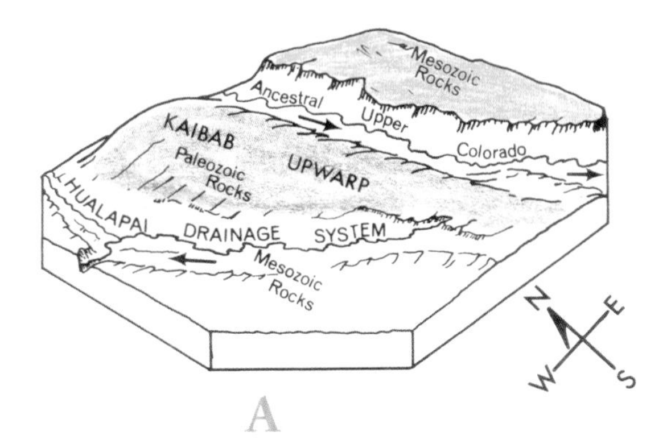

they called the ancestral upper Colorado River which, according to them, flowed out of Utah, then southeast towards the Rio Grande along the present course of the Little Colorado River but opposite its flow today. This drainage configuration was required because something had to erode suspected lake deposits from southeast Utah, and because the Muddy Creek Formation lacked material originating from the upper Colorado River.

STAGE 4 The southeast flow of the ancestral upper Colorado was disrupted when uplift near the Arizona–New Mexico state line reversed flow towards the west. This created a lake in the vicinity of present day Winslow and Holbrook and deposited the Bidahochi Formation. At the same time (and taking a cue from Charlie Hunt), the authors proposed that the Hualapai drainage may have flowed south in Peach Springs Wash. Some drainage may have gone west into the Muddy Creek basin but it didn't leave any sediment with rock types from the Grand Canyon. They stated that headward erosion by the Hualapai drainage back into the west side of the Kaibab Plateau may have already begun by this time.

STAGE 5 The Hualapai drainage extended its reach far enough east to capture the ancestral upper Colorado drainage and the through-going Colorado River was born. The capture point was defined as being near the confluence of the Colorado and Little Colorado rivers, east of the Kaibab upwarp. The time of this capture event was placed between about 10.6 to 2.6 million years ago. Modern dating

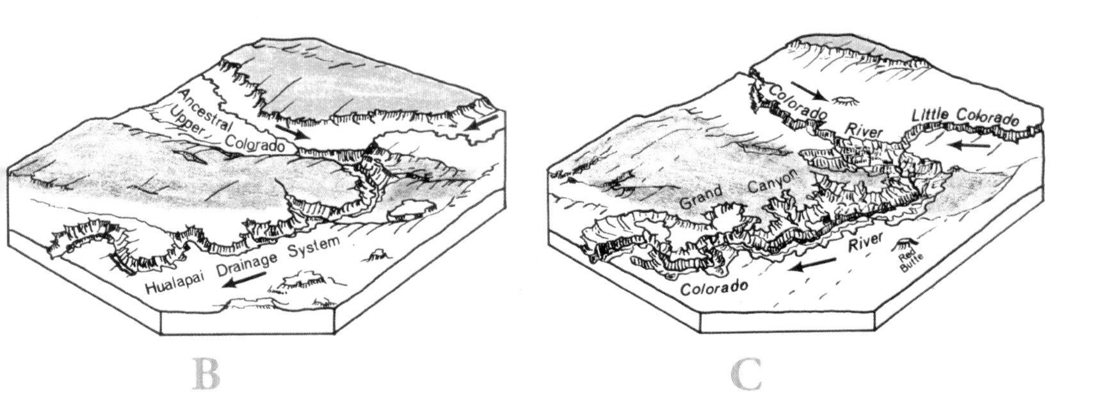

techniques have greatly refined these ages to between 5.8 and 4.4 million years ago. The dates were derived from the age of two lava flows—one from near Hoover Dam that covers pre-Colorado River deposits (5.8 million years) and one at Sandy Point on upper Lake Mead that covers unequivocal river deposits (4.4 million years). This is a good piece of detective work as the dates correlate well with other known events such as the timing of the opening of the Gulf of California.

The idea that the Colorado River could have been cobbled together from the prior existence of two separate and distinct river systems was a major advance not only in the understanding of the origin of the Grand Canyon but for the science of landscape evolution as a whole. In our modern world, it may often seem that what we accept as "facts" have always been with us. Yet ideas must be conceived and hatched for the first time by someone who receives a spark of insight for an original idea.

Although some major objections were raised within a few years after the publication of Bulletin #44, especially about a Rio Grande connection for the ancestral upper Colorado River, the results from this symposium mark a watershed point in our understanding of the evolution of the river. From this point forward geologists were not limited to a history that involved simply an antecedent, superposed, or subsequent origin. The river could now be viewed as an evolving entity with an equivocal beginning depending upon how it was defined. In a very real way, the floodgates were opened to many new and innovative ideas.

Charlie Hunt, 1969

In 1969 the U. S. Geological Survey commemorated the 100th anniversary of John Wesley Powell's first river trip through the Grand Canyon by publishing "Professional Paper 669." Charlie Hunt wrote the major article within it, "The Geological History of the Colorado River." This was a monumental work that examined the river's history from its headwaters to its mouth. In the thirteen years since his previous contribution, Hunt's own thinking regarding the origin of the Grand Canyon had evolved.

Hunt presented new ideas about how the Colorado River and its major tributaries may have altered their placement and configuration. He wrote that the upper Colorado may have terminated in a lake basin in the Glen Canyon region about 25 million years ago. Inflow did not exceed evaporation rates and therefore there was no outlet from this lake. According to him, it was the San Juan River that drained across what is now the Marble Canyon and Kaibab upwarp areas. He postulated that the Little Colorado River flowed farther south than it does today and that it and the ancient San Juan may have joined at some point west of the Kaibab Plateau. Given these vastly different orientations it's a wonder that he managed to retain the modern names for these ancient rivers! Sometime before 5 or 6 million years ago, the upper Colorado River overflowed the lake basin in Utah and joined these other two streams to create something akin to the modern Colorado River. These interpretations were novel, some might even say contorted, since few if any future workers followed up on them to any degree. Arriving finally downstream at the "Muddy Creek problem," (as all Grand Canyon geologists eventually must) Hunt stated:

I postulate that the ancestral Colorado River (that is, the ancestral San Juan and Little Colorado) left the Colorado Plateau via the dry canyon at Peach Springs.

This was an interesting idea resurrected in part from his 1956 paper. Geologists at this time were intrigued by this possibility but since it was first proposed, no one had actually found any evidence for Colorado River gravel in Peach Springs Wash. To ward off this criticism, Hunt noted that the river's pre-dam sediment load contained only 5 percent of rock types eroded from upstream of the

Grand Canyon. To him, this could explain why no evidence was found for the river going south in Peach Springs Wash. But couldn't that also explain the lack of evidence of a Colorado River in Muddy Creek time? It seems that Hunt was searching desperately to accommodate the ever strengthening view that the river could not have been flowing into the Muddy Creek basin when it was being filled with sediment between 16 and 6 million years.

Instead, Hunt introduced a seemingly outrageous hypothesis for the lack of any recognizable Colorado River gravels in the Muddy Creek Formation. He suggested that after leaving Peach Springs Wash to the south, the river became ponded and subsequently percolated into the subsurface. It dissolved limestone and this calcium-rich water was eventually piped northwest underground to springs that issued from the walls of the Grand Wash Cliffs. This water, according to Hunt, fed Hualapai Lake and deposited the Hualapai Limestone.

As a possible explanation for the lack of recognizable Colorado River gravels in the Muddy Creek basin, Hunt postulated that subsurface water draining to springs in the Grand Wash Cliffs (background) may have fed a lake that deposited the Hualapai Limestone (right foreground). Photograph by Wayne Ranney

Water . . . could have discharged through the cavernous limestone to supply springs . . . [at] the mouth of the Grand Canyon where the maximum deposition of the Hualapai Limestone is centered. Such a mechanism involves piping on a truly grand scale. Some may believe the scale is outrageous, yet . . . the postulated piping would provide the large quantity of water required for the lake and the limestone deposited within it and would explain the absence of [Colorado River] deposits in the lake.

Hunt was deeply troubled by the "Muddy Creek problem," since he saw evidence for the river being older upstream from the Grand Canyon in Colorado and Utah. He was equally ambivalent about a polyphase history for the river as proposed at the MNA symposium five years earlier. He believed that it was much easier to explain perceived age differences for the river by invoking the existence of lakes north of the Kaibab Plateau, even though deposits for such a scenario are lacking. The idea that the Colorado River experienced significant ponding events aligned him in many ways with the earlier views of Blackwelder but the two did not agree on the timing of these lakes. Surprising to many, Hunt's underground piping idea is being revived in some current ideas involving the Muddy Creek problem but his scenario for the Colorado River exiting through Peach Springs Wash appears for now to have died a quiet death.

Ivo Lucchitta and Richard Young

During the 1970s and 1980s, two of the youngest geologists to attend the 1964 symposium (as observers only and not among the twenty official participants), added greatly to our knowledge of how the Colorado River may have evolved. Both had been recruited by Eddie McKee to study gravel deposits in the area of western Grand Canyon. Ivo Lucchitta mapped the Muddy Creek Formation west of the Grand Wash Cliffs and Richard Young studied the relatively hidden gravels deposited within Milkweed Canyon, a tributary of the Colorado. Many of the known details about these two areas are the result of the more than thirty professional articles that these two geologists have written since then.

In modern times, Lucchitta has been the most articulate spokesperson on the history of the river in Grand Canyon and the Basin and Range. His views are unequivocally planted among those who argue for a "young" river but one that was integrated by headward erosion from two separate older systems. His observations concurred with Longwell about the Muddy Creek Formation being a pre-Colorado River interior basin deposit. Lucchitta also popularized the earlier findings of Babenroth and Strahler, which stated that the river probably established its course across the Kaibab Plateau in a subsequent valley formed between the crest of the upwarp and a southward retreating cliff of Mesozoic formations. This idea has become known as the "racetrack theory" since the subsequent valley mimicked the way a racetrack is confined between the grandstand wall and the infield of a racing track. Lucchitta believed that after leaving the area of the Kaibab upwarp this drainage went northwest away from the Grand Canyon region. He proposed that headward erosion by the Hualapai drainage captured the ancestral upper Colorado River west of the Kaibab upwarp in the vicinity of Kanab Creek. This was a variation to the findings from the 1964 symposium which stated that stream capture occurred east of the upwarp, near the confluence of the Colorado and Little Colorado rivers.

Young's work verified and expanded upon ideas related to the northeast drainage pattern that existed on the southern Colorado Plateau before 16 million years ago. This included the recognition of humid versus arid erosional styles, and the resulting landforms of plateaus and deep canyons. He documented the "unroofing" history of the Mogollon Highlands, whereby the uppermost, youngest rocks in the mountains were eroded first and consequently deposited as gravels at the bottom and oldest part of an ancient river deposit. His work helped determine that the "Rim gravels" are early Tertiary in age rather than mid-Tertiary as reported in the 1964 symposium results. Geologists such as Ivo Lucchitta and Richard Young are still active in Grand Canyon research topics and Richard Young served as chairperson for the symposium held at Grand Canyon Village in 2000.

Earl Lovejoy, 1980

By now it should be evident that there are essentially two key sticking points in establishing a coherent theory on the evolution of the Colorado River—the date of the "birth" of the river and the timing of the uplift of the Colorado Plateau. We cannot say when the river was born until we define what it is, but for now let's define it strictly as the river we see today fully integrated and flowing westward in its present course across the Kaibab upwarp towards the Basin and Range. Described in this way, the presence of the Muddy Creek Formation, which does not contain recognizable Colorado River sediment and is dated between 16 and 6 million years, means that the Colorado River could not have been "born" before 6 million years ago. Or could it?

Earl Lovejoy believed that the river could be older than this and still exiting the Grand Wash Cliffs during Muddy Creek deposition. Ironically, he used the very same Muddy Creek deposits that Longwell and Lucchitta had used to preclude the presence of the river in this locality! Lovejoy used by analogy an example from Big Bend National Park in Texas where the course of the Rio Grande, after exiting Santa Elena Canyon, is forced to flow against the base of an escarpment by the progressive growth of a large fan of flood deposits that enter the river from the opposite bank. Lovejoy invoked a similar setting for the Colorado River during Muddy Creek deposition, saying that the river's course may have been turned abruptly 90 degrees to the south immediately after emerging from the Grand Wash Cliffs. Such a sharp deflection, Lovejoy reasoned, could be accomplished by the progressive accumulation from west to east of Muddy Creek sediment towards the Grand Wash Cliffs. This, he explained, may have confined the river to a course narrowly set against the base of the cliffs and could also reasonably explain the lack of definitive Colorado River sediment within the bulk of the Muddy Creek Formation.

If this scenario were true then, according to Lovejoy, perhaps recognizable Colorado River sediments should be found exactly where the river is suspected to have flowed along the base of the southern Grand Wash Cliffs. Interestingly, there are significant sand and silt deposits within the Muddy Creek Formation in this position.

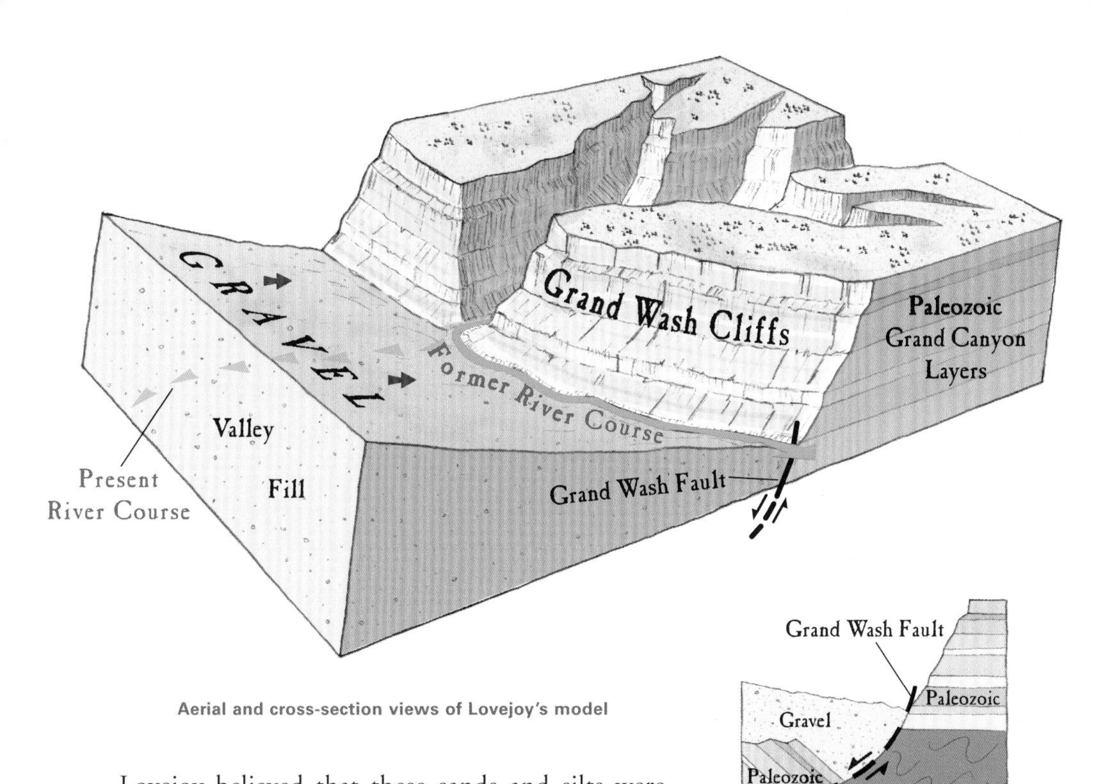

Aerial and cross-section views of Lovejoy's model

Lovejoy believed that these sands and silts were indistinguishable from modern Colorado River sediment. Concerning the admitted absence of more diagnostic Colorado River gravel, he used an example from the Truckee River near Reno, Nevada. There, diagnostic river gravel was present in an upstream reach of that river but settled and removed from the river's load above a narrow canyon, such that only sand and silt were deposited downstream in that river.

Lovejoy further suggested that the Colorado may have become ponded in the Hualapai Basin, where the rising lake water deposited the Hualapai Limestone on top of the coarser deposits of the Muddy Creek Formation. As this basin progressively filled with water he noted that the lake probably extended far up into an existing Grand Canyon. According to him all of the coarser debris in the river was deposited at the head of this lake far upstream in Grand Canyon. In this way, only clear water exited the Grand Wash Cliffs explaining the origin of the Hualapai Limestone. Eventually, as this basin became filled with algae-derived limestone, spillover of the lake established the modern course of the river to the west towards the Hoover Dam area. Ignoring the more dominant views of the time,

Lovejoy established a viable alternative to solving the "Muddy Creek problem."

Obviously, Lovejoy greatly favored the ideas of Powell, Dutton, and Hunt who envisioned an "old" Colorado River flowing west and was less influenced by the views of Blackwelder, Longwell, and Lucchitta who thought the river "young." For his theory to be valid however, he had to invoke an older age for the Basin and Range because the river could only flow to the west if a lowland existed there. The accumulating evidence however, suggests that highlands were present to the west of Grand Canyon until at least 30 million years ago and perhaps as recently as 16 million years ago and these highlands directed river flow to the northeast. Is it likely, as Lovejoy postulates, that an "old" Colorado River went west in early Tertiary time? Given what we know today, it is highly unlikely.

Ignoring Lovejoy's arguments for the timing of all of this and the defense for an old river, his observations regarding how the Colorado River may have exited the Grand Wash Cliffs during Muddy Creek time is an intriguing idea. It reveals to us the many ways by which geologists can circumvent seemingly irrefutable evidence to the contrary. It is no wonder that the story of the Colorado River remains full of enigmas and contradictions.

Don Elston

As we near the end of our survey of twentieth century ideas, one final concept deserves further elaboration—the notion of an old Grand Canyon. This idea has been championed foremost by Don Elston, retired geologist of the U. S. Geological Survey. Elston arrived on the Colorado Plateau in the 1950s to map uranium deposits, which were used in the effort to win the Cold War. However, his professional association with Charlie Hunt soon had him thinking about ideas on the origin of the Colorado River.

Elston saw evidence in the landscape for a period of deep dissection that may have commenced as early as 100 million years ago when, according to him, the plateau was tilted down towards the north. He believed (and continues to believe) that the Grand Canyon was cut to its present depth between 70 and 60 million years ago by a river that flowed to the northeast. Later, the canyon was completely buried by the "Rim gravels" as they aggraded and spread north from the Mogollon Highlands towards Utah and Wyoming. Subsequently, as the "Rim gravels" were eroded, the Grand Canyon was exhumed from beneath them. Then the modern Colorado River was established as the plateau reversed its tilt, (as William Morris Davis proposed in 1900) this time down to the south, causing the river to occupy the channel that previously went the other way. Important to this idea is the notion that the Grand Canyon was carved to its present depth between 70 and 60 million years ago.

Elston states that by about 10 million years ago arid conditions caused the Colorado River to become seasonal and that its channel became choked with local slope deposits to a depth of about 1000 feet. The "damming" of the river by these gravels could explain the lack of Colorado River deposits within the Muddy Creek Formation. According to him, the river flowed in the subsurface beneath the gravels through Grand Canyon towards the Grand Wash Cliffs. Wetter conditions later re-excavated the river channel.

The idea of an early Tertiary origin for the Grand Canyon has not been embraced by many others who insist that the river can be no older than 6 million years within canyon. But even though the idea of an "old" canyon does not represent a majority view, it does help us to define the outer limits for when the river could have evolved to its present configuration. We begin to see that deciphering the history of the Grand Canyon is similar to the story of how the three blind men describe the elephant that they can touch but cannot see. Each speaks the truth for that part of the animal that they happen to touch, but their descriptions sound as if they are describing three different animals.

The twentieth century closed without a widely accepted theory on Grand Canyon's origin. Older ideas were occasionally rehashed and reworked but newer interpretations were lacking during the last twenty years of the century. All of that was about to change however, when geologists began to yearn for another conference that would bring them together to consider the origins of the Colorado River and the Grand Canyon.

Early Twenty-First Century

We now enter a new millennium. The most current theories regarding Grand Canyon's origin build upon older ideas but use increasingly sophisticated techniques to glean more evidence from the landscape. New, more accurate dating techniques have been developed. The Geographical Information System allows for images from space to be digitized and viewed with precision like never before. A new era of active and changing thoughts regarding Grand Canyon's origin has begun.

Results of the 2000 Symposium

This new era of interest began in June of 2000, when a conference was held at Grand Canyon National Park. It was called "A Symposium on the Origin of the Colorado River." Prior to this time the only other conference ever convened to understand the river's puzzling evolution was the 1964 symposium held at the Museum of Northern Arizona—before the development of plate tectonic theory and at a time when many of the region-wide landscape relationships were not yet fully understood. Nevertheless a significant idea did emerge, namely that the Colorado River may have formed by the integration of separate and distinct pre-existing rivers. Although some of the details of this idea were later proved untenable, the larger contributions from the symposium remain as a significant milestone in our understanding of the river's possible evolution.

Colorado River: Origin and Evolution, 2004

About seventy geologists attended the 2000 symposium with thirty-six scientific abstracts submitted for presentation. Most authors presented their ideas orally in sessions. Two two-day field trips were organized. The pre-meeting field trip visited the Bidahochi Formation north of Holbrook, Arizona, and the Mogollon Rim area near Clint's Well. The post-meeting trip visited Milkweed Canyon on the Hualapai Plateau, and the Grand Wash

Cliffs area near Lake Mead. In 2004 Grand Canyon Association published a scientific monograph titled *Colorado River: Origin and Evolution*, which included thirty-three articles written, for the most part, by symposium participants.

Summarizing the broad range of ideas presented at the 2000 conference is an inherently subjective endeavor since participants observed the conference through the filter of their own experiences. New ideas are oftentimes met with vigorous resistance as geologists tend to be skeptics first and converts only when the evidence becomes overwhelming. However one might interpret the results of the conference, new ideas did enter the debate and they reinvigorated interest regarding the origin of Grand Canyon. On June 7, 2000 (the day the conference opened), the *New York Times* published a lead article on the front page of the Science Tuesday section, signaling that there was a resurgence of interest in the Grand Canyon story.

Four broad categories may serve to summarize the most current theories that took center stage at the 2000 conference:
1) Headward erosion and stream piracy
2) Spillover theory or the catastrophic draining of ancient lakes
3) Reverse flow of the river in all or parts of Grand Canyon
4) Significant Ice Age deepening of the canyon.

During the symposium, the tension between the proponents of headward erosion and spillover theory was palpable. Those who adhered to headward erosion expected a challenge from spillover enthusiasts and both were prepared to debate the facts. Ironically, both processes imply that the modern river originated from two separate drainage systems that were integrated at some later time; their differences lie in the manner by which the integration occurred. Reverse flow for the Colorado River could apply to the entire river in Grand Canyon as it exists today or just selected portions of it; differences here depend on whose evidence is being invoked. Proponents of reversal for only parts of the system must invoke an integration event, while those who envision the entire river as being reversed in Grand Canyon do not. The last category addresses only how the canyon may have been deepened in the last few million years, perhaps in response to climate change or differential uplift along faults, or both.

Headward Erosion and Stream Piracy

In recent years, some geologists have questioned the ability of headward erosion to act upon arid-land river systems, especially in relation to the development of the Grand Canyon. They point out that in canyon country runoff in the rather limited headwall area of steep-gradient streams may be deficient or too ineffective to cause them to lengthen their courses significantly. At the beginning of the 2000 symposium, there was a buzz in the air that some geologists came prepared to bury forever ideas related to headward erosion. However, the old guard came to its defense both directly and tangentially; they were armed with some new twists on the idea and some fresh insights from younger scientists.

Headward erosion is most frequently thought to result from uplift (although climate change can also induce it) since uplift can increase the gradient of streams, strengthening their erosive power and enlarging their drainage area. But the timing of Colorado Plateau uplift remains controversial, also bringing into question the mechanism for headward erosion. Ivo Lucchitta and others argue for a significant amount of plateau uplift within the last 6 million years. They believe that this uplift has been the driving force in creating the Grand Canyon through headward erosion. The evidence is derived in part from deposits that are believed to have been laid down in the Gulf of California but are now as much as eighteen hundred feet above sea level. Some geologists reject the idea that these deposits are marine and use this as evidence against late Tertiary uplift. The timing and frequency of plateau uplift continues to be vigorously debated and with it the importance of headward erosion in creating Grand Canyon.

Ivo Lucchitta continues to argue for a Grand Canyon that is younger than 6 million years. He completed mapping the Shivwits Plateau in the early 1970s and found gravels beneath a 6 million year old lava flow that came from south of the Grand Canyon; his interpretation is that the Grand Canyon did not exist prior to this time. He implies that a mature plateau landscape dominated the Grand Canyon region until the cycle of canyon cutting began sometime about 6 million years ago. Lucchitta further states that the Vermilion Cliffs north of Grand Canyon were stripped from the

Shivwits Plateau only in the last 8 to 6 million years. This is a surprisingly young age considering that the humid conditions needed for lateral stripping were dominant only in the early Tertiary and that these cliffs are now located far to the north near Colorado City.

At the 2000 conference, one of Richard Young's graduate students demonstrated how West Clear Creek on the Mogollon Rim most likely has eroded headward into East Clear Creek capturing some of its drainage area in the process. The study was completed to show that irrefutable evidence exists for headward erosion acting upon the edge of the Colorado Plateau, even on a stream much smaller than the Colorado River. This headward erosion occurred within the last 10 million years and has progressed back into the plateau edge by more than twenty miles.

Jim Faulds and others have done extensive research on the faults that created the Grand Wash Cliffs at the west end of Grand Canyon and have obtained age constraints on when this may have occurred. They show that faulting began around 13 million years ago and suggest that this could be when headward erosion began cutting into the western edge of the plateau. This finding seems to refute one of the major arguments against headward erosion. The abrupt appearance of Colorado River gravel after 6 million years ago suggests to some geologists that headward erosion progressed rather instantaneously at least 150 miles upstream. Faulds and his colleagues state that headward erosion into the plateau edge could have operated over a much longer 7-million-year period, culminating in an integration event sometime between 5.8 and 4.4 million years ago.

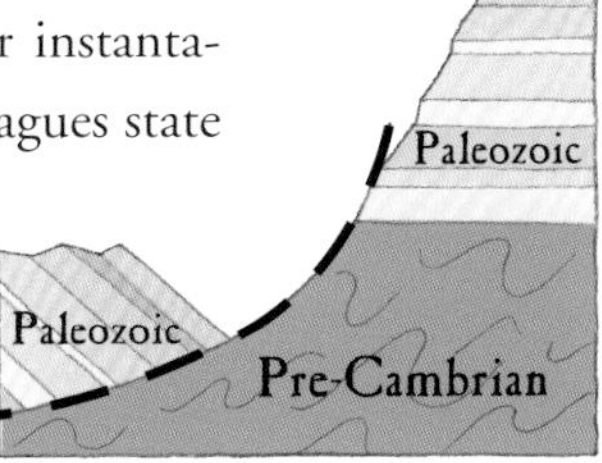

Faulting created the Grand Wash Cliffs which mark the boundary between the Colorado Plateau and Basin and Range provinces. Photograph by Wayne Ranney

Even though the efficiency of headward erosion has been questioned recently by some geologists, good evidence remains that it could have been an important process in creating the Grand Canyon. It is hard to imagine this process operating under the current arid climatic conditions or within the modern, deeply dissected landscape. However, former climates and former topography could have enhanced its ability to shape the landscape and it has been documented that past climate and topography was much different than it is today. Perhaps reports that headward erosion is "DOA" at Grand Canyon are greatly exaggerated.

Spillover Theory

If any one process discussed at the 2000 conference served to challenge the preeminent status of headward erosion, it was the idea that closed, interior basins on and adjacent to the Colorado Plateau could have filled with water and sequentially spilled to integrate the modern Colorado River. And even though some geologists remain doubtful that a large, long-lived Lake Bidahochi ever existed east of the Grand Canyon, others not only find evidence for this lake's existence but also for its catastrophic draining as well. Spillover theory has nudged its way to the forefront of ideas regarding the origin of the Colorado River.

In 1934, Eliot Blackwelder was the first geologist to suggest that the modern river was the result of closed-basin spillover along its length. He envisioned this setting without much sedimentary evidence for lakes; rather his ideas were formulated by recognizing the curious fact that today the river flows alternately through wide open basins (where the lakes would have been located) connected by narrow canyons (where these lakes overtopped and spilled). During the 1964 symposium, geologists postulated the existence of Lake Bidahochi along what is now the course of the Little Colorado River in the vicinity of Winslow and Holbrook, Arizona. At the 2000 conference, a pre-meeting field trip to this area resulted in mixed reviews about whether these deposits were laid down in an expansive lake as envisioned in 1964 or were merely the scattered remnants of disconnected, ephemeral ponds.

However, Norman Meek and John Douglass came to the defense of a large lake that spilled catastrophically to create the course of the Colorado through Grand Canyon. They used modern analogies from elsewhere in the Southwest and described how the Mojave River in California catastrophically overflowed basins along its path to establish its present course. Meek and Douglass proposed a similar scenario occurring after 6.5 million years ago when a suspected basin in central Utah overflowed and filled Lake Bidahochi. According to them, at about 5.5 million years ago, Lake Bidahochi overflowed, initiating the incision of the Grand Canyon, perhaps at a low spot in that basin near today's Grandview Point. Curiously, they arrived at this conclusion without being aware of Eliot Blackwelder's earlier hypothesis. They added that their overflow theory could explain how a deep canyon was carved even though evidence continues to mount against recent plateau uplift. Some geologists have said, "No uplift, no canyons," but Meek and Douglas modified this line of thinking to state, "No base level change, no canyons." They refer to the idea that spillover of Lake Bidahochi would have been directed into the topographically low Basin and Range and this immediate connection between the two only appears in the evidence like recent uplift.

Meek and Douglass did not specify whether the catastrophic draining of Lake Bidahochi created its own line of drainage or whether it followed some preexisting channel. On these grounds, some geologists have tried to marginalize or even dismiss their idea. However, it may be possible that their proposed spillover might have flowed into the drainage suggested by Jim Faulds and his colleagues, who thought that a headward eroding stream may have begun 13 million years ago when faulting began creating the Grand Wash Cliffs. It is tantalizing to combine these two ideas since it is plausible and, somewhat ironically, incorporates both headward erosion and catastrophic spillover as coeval mechanisms for creating the modern river.

Jon Spencer and Philip Pearthree offered a partial integration of these ideas. They provided a thoughtful critique on the inefficiency of rapid headward erosion in the arid southwest and proposed that closed-basin spillover was a more likely process to integrate the

Colorado River through Grand Canyon. They, like Blackwelder before them, proposed that the Colorado River may have flowed from its lofty headwaters into lakes located on the central Colorado Plateau about 15 million years ago. They note that the rate of sediment accumulation then was only one third of present-day rates and this could explain the lack of evidence for such lakes. They stated that when one basin overflowed, it could have catastrophically filled the next basin, causing it to catastrophically fill the next, and so on downstream. They speculated that Lake Bidahochi could have filled rapidly in such a way and then spilled to the west to carve a course for the river through Grand Canyon. A short-lived Lake Bidahochi would not leave much evidence for its brief existence. Their postulations were meant as an alternative to the idea that the river was integrated by headward erosion on the west side of the Kaibab Plateau.

Jonathan Patchett and Jon Spencer provided evidence for a possible spillover origin of the lower Colorado River in the Mojave Desert. In this location the river flows through three wide basins separated by narrow canyons which are cut into mountainous bedrock. Within these broad basins are isolated outcrops of the Bouse Formation, found at various elevations as high as eighteen hundred feet above sea level. Some geologists explain the separation of these Bouse outcrops as originating from a late-stage uplift of the plateau edge across formerly continuous marine deposits of the Gulf of California. But Patchett and Spencer note that no faults have been found between the outcrops and suggested instead that they were deposited in separate freshwater basins that were integrated by closed-basin spillover.

The environment of deposition for the Bouse Formation remains highly controversial but possible resolution of the dilemma has great implications for proponents of headward erosion versus those for closed-basin spillover. If the Bouse deposits are marine in origin, the present day outcrop pattern necessitates a period of recent plateau uplift. If it formed in a freshwater environment, then recent plateau uplift is unnecessary. The fossil evidence, as well as the chemistry of certain elements within the Bouse, seems to support interpretations for either a freshwater or a marine environment. This makes the debate confusing at best. Patchett and

Spencer have offered a plausible explanation that fits well with other scenarios for closed-basin spillover elsewhere along the river.

Bob Scarborough, another proponent of spillover theory, thinks a pre-spillover canyon was cut during the period of northeast drainage but only as deep as the Esplanade level in western Grand Canyon. He envisions the pre-spillover river as a highly meandering stream on the Esplanade surface that could have helped carve the spectacular temples and the Chuar Valley found in the eastern part of Grand Canyon. According to him, the spillover of Lake Bidahochi towards the west followed a channel cut originally during the time of northeast drainage. Scarborough believes this new stream may have had to clear rocky debris that choked the old channel but ultimately it began to dissect the modern Grand Canyon.

Spillover theory appears to be gaining wider acceptance among geologists since the 2000 conference. Although unequivocal lake deposits on the Colorado Plateau are lacking, other evidence from the Basin and Range supports the idea or at least strengthens the argument against headward erosion. These two processes are now running neck and neck as the two most likely candidates for how the Colorado River originated in Grand Canyon.

Reversed Flow of Colorado River

The idea that the Colorado River may have reversed its direction of flow is one of the more interesting aspects concerning the river's possible evolution. Overwhelming evidence supports the idea that the Colorado River, or some precursor to it, flowed towards the northeast in the early Tertiary, opposite its flow today. This evidence requires that a period of drainage reversal occurred in the area but the extent to which the modern landscape preserves elements of this ancient drainage system remains unknown, controversial, and enticing to speculate upon.

Richard Young was the first geologist to widely publish information on gravels preserved in western Grand Canyon from streams that flowed northeast. He has mapped and described a nearly-twelve-hundred-foot section of these deposits in Milkweed Canyon on the Hualapai Indian Reservation. These rocks and the deep

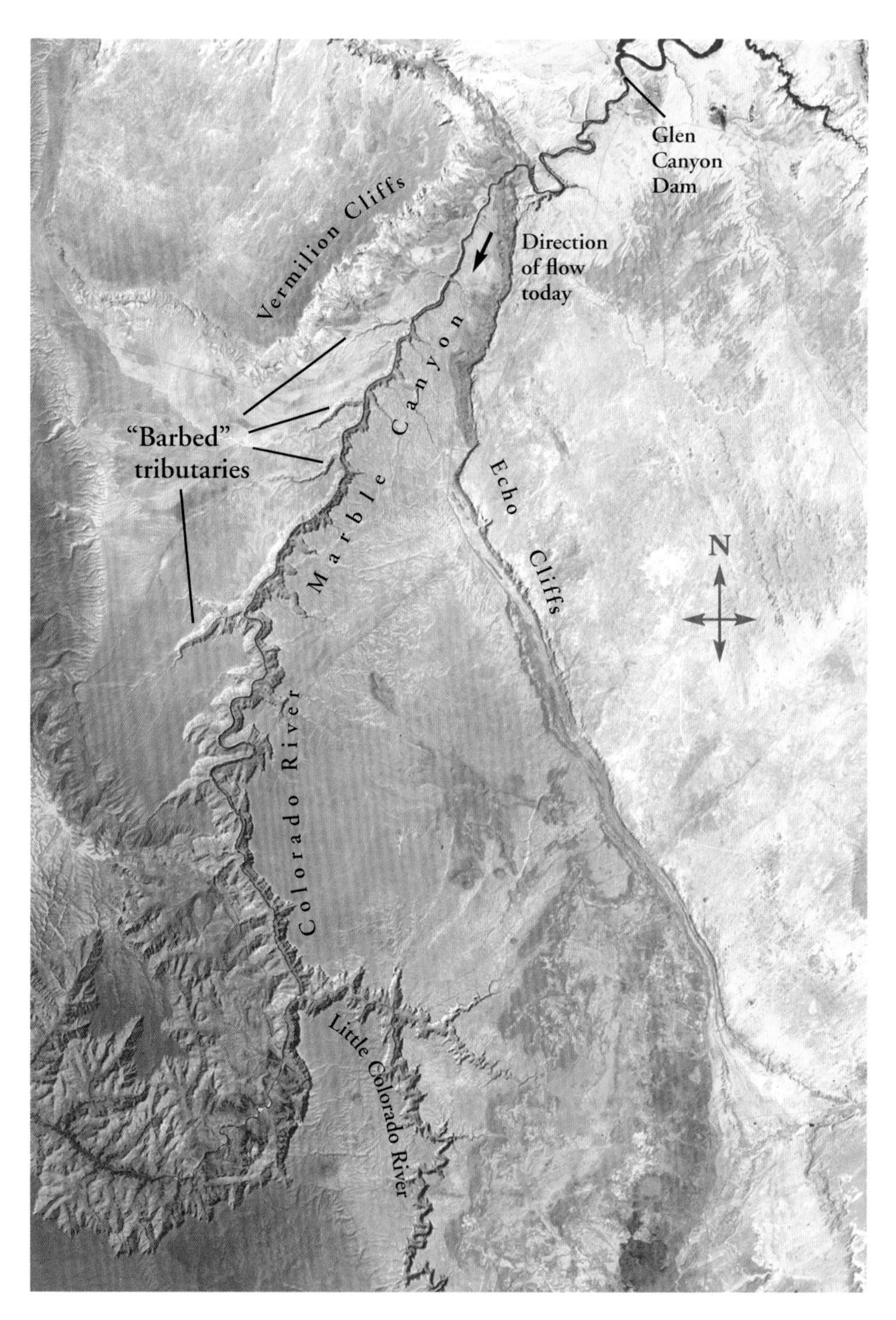

"Barbed" tributaries of Marble Canyon can be seen in this Landsat image. Typically, tributaries enter the master stream pointing in the downstream direction, suggesting that this section of the Colorado River may have reversed its direction of flow. Landsat image courtesy of U.S. Geological Survey, Southwest Geographic Science Team, Flagstaff Office canyon in which they are exposed, display evidence for the progressive "unroofing" of the Mogollon Highlands, show how the climate became more arid during the course of the Tertiary, and suggest that uplift of at least four thousand feet was accomplished in western Grand Canyon during the early Tertiary. Young has also proposed that at least three hundred feet of Tertiary sediment may have covered the western Grand Canyon area and that superposition on this surface may have emplaced the Colorado River in its present course there.

Andre Potochnik has studied the similarities between an ancestral Salt River Canyon in central Arizona and the Grand Canyon. The Salt River paleocanyon was cut to a depth of between three thousand and forty-five hundred feet during the period of northeast flow. Drainage reversal occurred there sometime after 14.8 million years ago; the evidence is based on flow directions preserved in various gravel deposits. Potochnik states that the Colorado River's history in Grand Canyon may be analogous to what has been documented in the Salt River paleocanyon and that the course of the Colorado River today is a relic from when the river flowed northeast. Don Elston, you will recall, accepts this broad interpretation as well, with the added condition that the present depth of the canyon was attained during the period of northeast flow. The 14.8 million year date for drainage reversal in the Salt River paleocanyon agrees in general with Jim Faulds's interpretation of when headward erosion may have begun eating its way east into the edge of the Grand Wash Cliffs.

The author of this book presented an idea at the conference that the Marble Canyon section of the Colorado River may be an inherited relic of northeast drainage and suggested a place where an integration event may have occurred. The evidence is seen in both the "barbed tributaries" in Marble Canyon and the river's relationship to the Butte Fault upstream and downstream of the confluence with the Little Colorado River. According to this idea, the Marble Canyon section of the Colorado and its tributaries were positioned by northeast flowing streams, consequent to the northeast dip of the strata and most likely within Mesozoic rocks before they were stripped back to create the present day Echo and Vermilion cliffs. This drainage was possibly fed by the ancestral Little Colorado

River which, itself, originated as a northward-directed stream flowing out of the Mogollon Highlands.

Further support for the specific location of the integration of separate systems at the confluence of the Colorado and Little Colorado rivers comes from the curious relationship between the Butte Fault and the Colorado River. Recall that in 1890, C. D. Walcott reported on the parallel relationship between the fault and the river in the ten-mile stretch between Nankoweap Canyon and the confluence with the Little Colorado River. However, downstream from the confluence, the Colorado seems to ignore the fault; it approaches and soon crosses it, creating the Big Bend in the river below Desert View. Is it likely that these differing relationships of the river to the fault above and below the confluence are only a coincidence? Knowing that northeast drainage was once present in the region and given the likelihood of some integration event to create the modern Colorado, a more likely hypothesis is that these two sections of the river have separate histories. The idea agrees with many other aspects of the known history of the river. Although there is no hard evidence such as gravels, it invokes the period of northeast drainage, is supported by the accepted fact that the Butte Fault is older than the river and thus influenced the position of the river, and offers a specific location for an integration event, either by headward erosion or catastrophic spillover.

An important question remains however. Does the Colorado River in Grand Canyon preserve evidence in the modern landscape of the early northeast drainage systems? Increasingly it seems that the answer to that is, Yes.

Ice Age Deepening of the Grand Canyon

The previous ideas originated as older theories that were refined, updated, and reinvigorated by fresh attention and new debate at the 2000 conference. The recognition of Ice Age deepening in the Grand Canyon however, is a relatively new concept that has excited the scientific community much like the polyphase history for the river did at the end of the 1964 symposium. The Ice Age history of the Colorado River shines a bright light on specific processes that may have deepened and widened the canyon and the rates at which this could have been accomplished. The role that glacial meltwater may have played in deepening the canyon was long ignored by scientists. Along with the new recognition that Ice Age tectonic events altered the relative elevations of the western and eastern parts of the canyon, these two concepts have moved to the forefront of current theories.

Ice did not carve the Grand Canyon or affect its shape or profile in any direct way; the glaciers were never this far south. But the Colorado River extends its reach far north into landscapes that were glaciated. This means that glacial meltwater traveled through the Grand Canyon on many occasions during the last two million years. The Ice Age was not a static environment where ice appeared and melted only once, rather it was a time when mountain glaciers advanced and retreated numerous times, alternately storing and freeing large amounts of water relatively quickly. Actually, the earth is still in the Ice Age; the last 10,000 years are defined as an interglacial, a period of warmer temperatures and altered precipitation patterns that separate glacial advances from intervening retreats.

Historically, very little attention has been directed towards the role that climate may have had played in carving Grand Canyon. Only one climatologist attended the 2000 symposium and he directed his observations to the period of time when an integration event may have occurred, 3 to 4 million years before the onset of the last Ice Age. But as shown in Chapter 3, increased runoff in rivers creates a greater capacity to carve canyons. With increased precipitation in the Rocky Mountains and on the Colorado Plateau, mountain glaciers repeatedly released their stored water and it is easy to envision how the canyon could have been significantly deepened during this time. This is especially true if climatic factors were given a boost by tectonic ones.

Increased cutting capacity of the Colorado River by Ice Age runoff could have been enhanced by movement on the Toroweap and Hurricane faults in western Grand Canyon based on studies completed by Cassie Fenton, Bob Webb, and Philip Pearthree. The faults broke volcanic rocks and alluvial fans that are between 30,000 and 400,000 years old. These geologists then measured the total displacement of the two faults at about nineteen hundred feet and extrapolated that the bulk of this displacement could have occurred in the last 3.5 million years. Their idea suggests that as the western Grand Canyon was lowered, cutting of the eastern canyon accelerated.

Joel Pederson and Karl Karlstrom looked at the rates of displacement on the same two faults but concluded that these rates were insufficient to carve the present depth of the entire Grand Canyon. They believe that the displacement rate on the Toroweap Fault is not large enough to drive upstream deepening of the canyon and that the earlier integration event, at 6 million years, is the more important driver in deepening Grand Canyon. Their conclusion states that the rate of canyon cutting has either diminished through the last 6 million years (as the river progressively lowered its gradient) or alternatively, some of the canyon's present depth was achieved before integration of the river took place.

With more refined dating techniques, geologists have determined new ages for Grand Canyon lava flows which show them to be about one third the age previously thought. The oldest flows are now known to be on the order of 630,000 years. These more refined dates allow geologists to determine the rate at which lava dams were removed and replaced by later dams. These studies help in determining how fast the Colorado River can remove an obstacle from its path. With lava dams now more tightly constrained in time, relatively quick removal of these obstacles is implied. Carol Hill and Victor Polyak of the University of New Mexico are presently obtaining dates on deposits in caves and mines that may help to determine the rate of incision of the Colorado River and thus the age of the Grand Canyon. Certain cave formations such as stalactites, stalagmites, spar linings, mammillaries, and some gypsum deposits form either below, at, or above the water table. By dating these kinds of deposits, the position of the water table over time in various parts of the Grand Canyon can be ascertained. These caves and mines are located primarily within the Redwall Limestone and whatever dates are obtained will shed light on when the water table, and by inference the river, was at this level.

Noel Eberz has been fascinated by the presence of the many caves in the Redwall Limestone in certain sections of Marble

Canyon. He believes that these caverns document the importance of groundwater processes in shaping the course of the Colorado River in eastern Grand Canyon. The idea is that stream capture of the ancestral upper Colorado in Marble Canyon was accomplished by subterranean piping of groundwater through a cavern system. Eventually, this cavern system collapsed and the river channel between Nankoweap and Grandview Point was born. This suggests that the river's flow across the Kaibab upwarp was established first in the subsurface, but cavern collapse caused the drainage to flow on the surface at about 6 million years ago. This could readily explain the problem of how the river crossed the upwarp.

Finally, Thomas Hanks, Ivo Lucchitta, and others looked at gravel surfaces north of Navajo Mountain and reported on their subsequent dissection. This study suggests that Glen Canyon, at over eight hundred feet deep, was cut within the last 500,000 years. All of these studies show that Ice Age deepening of the Grand Canyon was significant, on the order of two thousand feet or more.

SUMMARY

The picture that is emerging after the 2000 symposium is of northeast drainage across northern Arizona from about 80 to perhaps 16 million years ago. Some of these early drainages were cutting into Grand Canyon stratigraphy on the Hualapai Plateau. Perhaps this was an "old" Grand Canyon, but it was a much different river system than the one we see today. Headward erosion from the west may have traveled eastward into this pre-existing system from the Grand Wash Cliffs. A catastrophic spillover of lakes connecting the east with the west cannot be discounted. Doubling of the depth of the canyon has occurred in very recent times.

5

Landscape Evolution of the Grand Canyon Region

Grand Canyon's Inner Gorge is the deepest point on the Colorado Plateau landscape. Photography by Wayne Ranney

The precise sequence of events that formed the Grand Canyon is unknown. However, a few well documented geologic events allow us to paint a broad picture of how the canyon may have evolved. These events are the broad uplift of the region after withdrawal of the seas, the lowering of base level downstream from the Grand Canyon, and the ensuing deepening of the canyon by the river.

MANY PUZZLES REMAIN concerning the origin of the Grand Canyon. Very few clues have survived from its subtle beginnings into the modern era of deep landscape dissection. Nevertheless, geologists have solved some of the puzzles of its origin by looking at the tiny bits of evidence that remain scattered within the canyon, across the plateau, and in adjacent areas. If we use broad brushstrokes in forwarding an explanation, and understand that some of the details may never be known to us, we can postulate a generalized sequence of events that may have given rise to this unique landscape.

It was unknown to John Wesley Powell in 1875, when he proposed that the Green River in Utah was older than the landscape surrounding it, that the modern Colorado River system may be one of the younger landscape features present on the Colorado Plateau. The idea of a young Colorado River, however, can only be true if we accept the rather limited definition of the river as a fully integrated, southwest-flowing stream, that lies deep within the plateau landscape, for which all or part may be inherited from older and significantly altered river systems. In other words, we can only say how old the Colorado River might be, based on how we define what it is and what constitutes its beginning.

A hypothetical example may be in order. Say, that by some stroke of magic, we obtained a time-lapse film of the history of the Colorado River. With this film, which would have to cover at least 80 million years of Earth history, we would witness the evolution of Grand Canyon. Imagine, if the film was four hours long, every second would represent 5,555 years! A million years of Earth history would be represented by only three minutes of film. Now, if the movie were played backwards so that we watched the river evolve back towards its inception, how much change could we tolerate in the river's configuration, position, flow direction, or length and still feel comfortable using the name Colorado River? In such a film, we would start out with a river quite recognizable, but every frame shown in reverse would change or remove features of the modern system that could challenge our use of the name Colorado River. Some segments of the river might "disappear," while others might flow in the opposite direction. Would that still be the Colorado River? Perhaps now we can appreciate the difficulty in determining a precise age for the beginning of the river and Grand Canyon, and why geologists still disagree on whether the canyon may be old or young.

In attempting to unravel the mysteries of the Colorado River and the Grand Canyon, we must be aware that competing and seemingly opposed lines of reasoning can somehow lead to the same resulting landscape. And when we favor one idea at the expense of others, we are necessarily forced into a much narrower line of descent with fewer alternatives possible thereafter. In many ways, the Grand Canyon story is like a maze in that by accepting a viable idea at the start, we may be committed to following it through wherever it leads us. Some geologists prefer to look at the earliest history of the river and attempt to evolve it forward in time. Others emphasize the most recent evidence and attempt to work backwards in time. The results can be strikingly different and an outsider observing it all may wonder if these geologists are addressing the same problem at all. Many of the intimate details of the story remain far beyond our grasp.

In any case, it may be helpful to subdivide the known history of the Colorado River and Grand Canyon into discrete time divisions that have been recognized as being separate and distinct from the periods that precede or follow them. Six time divisions can be identified but not all are equally represented by the evidence. Some portions of the history must be guessed at because so few deposits exist from these time periods. However, using what is known from adjacent areas, a possible scenario can be postulated that would weave a plausible sequence of events. These are the following:

1) 540 to 70 million years ago: Accumulation of colorful, flat-lying strata
2) 70 to 30 million years ago: Uplift of the Mogollon Highlands and initial drainage towards the northeast
3) 30 to 16 million years ago: Initial collapse of the Mogollon Highlands and a "confused" drainage pattern
4) 16 to 6 million years ago: Creation of the Basin and Range Province and interior drainage
5) 6 to 2 million years ago: Evidence for the beginning of an integrated Colorado River flowing to the Gulf of California
6) 2 million years ago to the present: Ice Age deepening of the Grand Canyon

A light snowfall accentuates layers of sedimentary rock east of Grand Canyon Village. Photograph by Chuck Lawsen

540 to 70 Million years Ago: Accumulation of Grand Canyon Strata

Prior to the establishment of the Colorado River system, the Colorado Plateau was a low-lying depositional basin that accumulated a thick blanket of sedimentary rocks. The future plateau country looked nothing like it does today—the extensive flat-lying sedimentary rocks reveal that the landscape during this time was quite featureless for hundreds of miles in every direction. It was as unremarkable a landscape as the Mississippi River Delta is today. More than fifteen thousand feet of sediment accumulated during a 470-million-year time span, most of it deposited at or below sea level. The river environments that once existed in the future Grand Canyon area, as represented by the Hermit Formation, for example, were completely buried by younger deposits thus precluding any direct relationship between them and the Colorado River we see today. There are no river deposits from this huge time span that are connected in any way to the origin of the Colorado River.

70 to 30 Million Years Ago: Uplift of the Mogollon Highlands and Northeast Drainage

Rivers obviously cannot exist in areas submerged by the sea; consequently, the story of the Colorado River could not begin until the ocean withdrew from the area for the final time, which happened about 80 million years ago. By 70 million years ago uplift of the plateau and surrounding regions began; and sometime between 80 and 70 million years ago marks the hazy beginning of the Colorado River story. The retreating seaway exposed a low-lying coastal plain that was the blank canvas upon which the Grand Canyon eventually would be carved. Nascent river drainages were soon established on this virgin landscape and they most likely exerted some influence on the configuration and location of future drainage systems. This is

The Laramide Orogeny

The mountain building event that created the Mogollon Highlands is called the Laramide Orogeny. Geologists have long debated whether this orogeny, or mountain building event, began 80 or 70 million years ago but the 70-million-year-old date seems to be gaining favor, at least with a vocal few. It lasted until about 40 million years ago. The Laramide created not only the Mogollon Highlands but also raised, relative to surrounding areas, the Rocky Mountains, the Kaibab upwarp, and all of the monoclines of the Colorado Plateau. Most of the Laramide Orogeny occurred within a time period known as the Tertiary, which lasted from 65 to 2 million years ago. The Tertiary is an important period in Grand Canyon's history and can be informally subdivided into early, middle, and late periods. The term "Tertiary" is used frequently in this book to refer to this expanse of time rather than frequently repeating specific numbers of years.

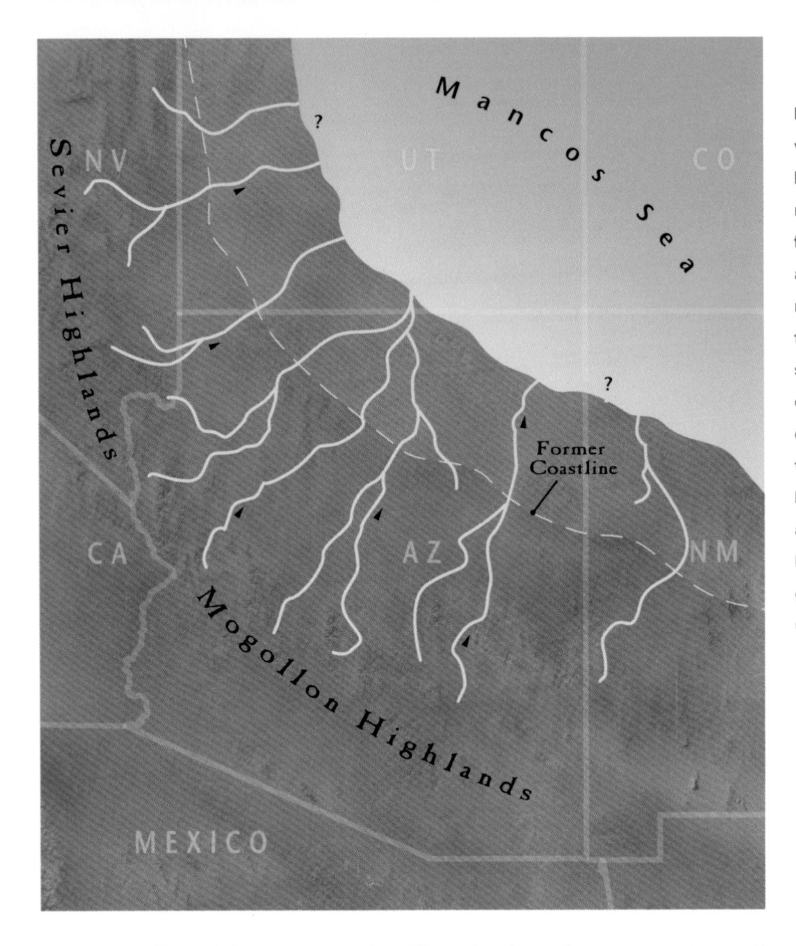

By 80 to 70 million years ago (Ma), the Mancos Sea receded northeast of the future Grand Canyon, and rivers flow north-east towards the retreating shoreline. Ironically, even though flow direction is opposite the modern Colorado River, it is the one aspect of the river's history that most geologists agree upon.

true even though later events significantly altered many portions of these early drainages.

The Mancos Shale is the deposit left behind by this final sea. Today, its nearest exposure to the Grand Canyon is just west of Glen Canyon Dam, but it is likely that it once extended farther south before erosion stripped it back to its present location. About 80 million years ago, this sea began retreating to the northeast and by 70 million years ago a large mountain range was formed to the southwest of the Grand Canyon area. Geologists call these mountains the Mogollon Highlands. The highlands were uplifted as the North American continent drifted west and collided with the oceanic Farallon Plate. Continued uplift of the Mogollon Highlands caused the sea to retreat farther away to the northeast. This caused the first drainages that formed on this emerging surface to flow from the mountainous southwest to the coastal northeast, *exactly opposite the flow of the modern Colorado River.*

Remnants of the Music Mountain Formation (center) deposited in Milkweed Canyon document that significant erosion into Grand Canyon strata had occurred before gravel deposition here. Photographs by Wayne Ranney

These ancient drainage patterns probably resembled what is seen today in the western part of the South American landscape. Colliding with the Nazca Plate, the Andes Mountains are experiencing active uplift and dynamic volcanism. From the mountain crest, rivers drain to the east and meander sluggishly into the low-lying Amazon Basin. The western Amazon Basin, located in Peru, Bolivia, and Brazil, is likely a perfect modern analog for how drainage was organized on the southern Colorado Plateau between about 70 and 30 million years ago. Like the Amazon Basin, the early plateau landscape had a moist, tropical climate which hastened rock weathering and influenced erosion styles. Like the Andes Mountains, the Mogollon Highlands experienced uplift and widespread volcanism.

The regional plate tectonic evidence for an initial northeast drainage pattern on the Colorado Plateau is further aided by sedimentary evidence found in a tributary of the Grand Canyon. Milkweed Canyon, located on the Hualapai Plateau west of Peach Springs, Arizona, contains remnant river gravels that document flow direction to the northeast away from the Mogollon Highlands. These deposits are called the Music Mountain Formation and are related to a larger group of river sediments that have been informally and historically called the "Rim gravels," since they are sporadically found on the top of the Mogollon Rim in central Arizona. They invariably show that drainage was directed to the northeast across the

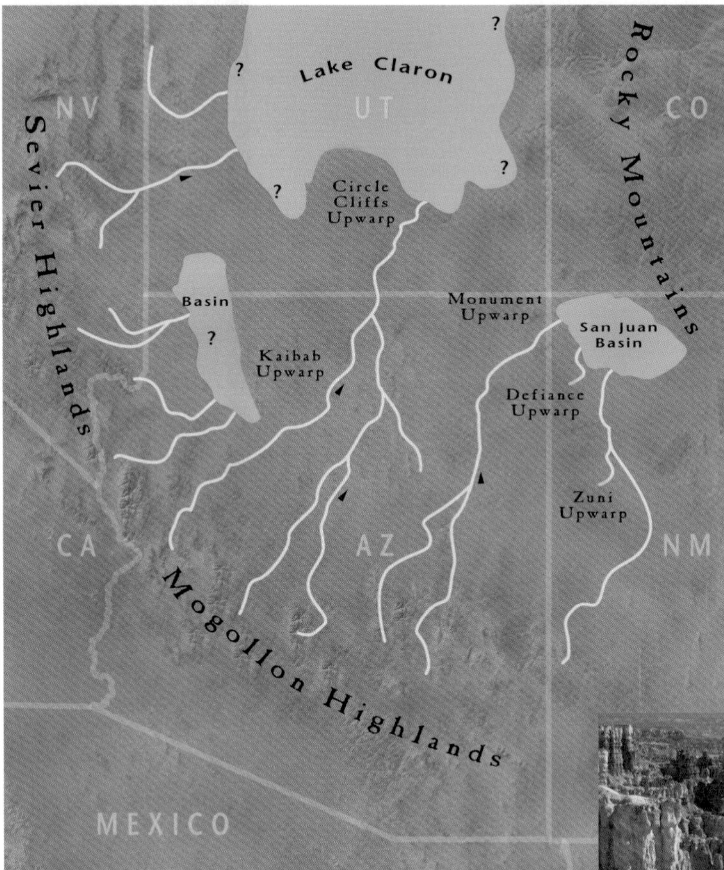

By 70 to 30 Ma, the Mogollon and Sevier highlands serve as headwaters for the northeast-flowing streams that drain to freshwater lakes on the future Colorado Plateau.

Strata in Bryce Canyon (below) represent the location of one of these lakes. Photograph by Bob and Suzanne Clemenz

Grand Canyon region. These streams probably flowed first into the retreating seaway, then, when the Laramide Orogeny created the Rocky Mountains, into fresh water lakes in southern Utah. Evidence for these lakes is preserved today in the delicate hoodoos in Bryce Canyon National Park.

These gravels contain well-rounded clasts of rock types exposed only to the southwest of the Grand Canyon area. The rounding shows that they traveled relatively far, while the rock type indicates that they could only have come from the southwest, since that is the only place where rocks of this type have been exposed to erosion. Taken together with the overwhelming evidence for the larger tectonic elements in the region, the "Rim gravels" and the Music Mountain Formation confirm that northeast drainage was present across the Grand Canyon region during this time. Paradoxically, this necessitates a later period of drainage reversal on the southern plateau and the idea of drainage reversal is one of the more confusing aspects of the whole Grand Canyon story. Some geologists believe that the "Rim gravels" may have completely blanketed the Grand Canyon area and that the eventual course of the Colorado River was set upon the swales of this surface. Others aren't so sure.

Precise age dating of these gravels has proved frustrating. Scant fossil evidence from nearby rocks on the Coconino Plateau and from the Bryce Canyon deposits suggests that they are about 40 million years old. On the White Mountain Apache Indian Reservation southeast of the Grand Canyon, gravel deposits that are also part of a northeast flowing drainage system have been dated at about 33 million years old. And northeast directed stream channels on top of the Grand Wash Cliffs, near Kingman, Arizona are filled with 18.5-million-year-old volcanic ash flows, suggesting an even more recent date for drainage in this direction. It is likely that drainage in northern Arizona was to the northeast from about 70 to 30 million years ago and maybe as recently as 16 million years ago.

As fascinating as this drainage scenario might be, one naturally asks whether portions of the Grand Canyon could have been carved by these drainages and if so, how deep? Again, we must define exactly what the Grand Canyon is before we can answer whether or not it originated at this time. For example, what if only a twenty-five-mile stretch of the modern Grand Canyon was carved by these streams? And what if that length was cut only one thousand feet deep or into strata that are now completely removed from the canyon. Would that qualify as the beginning of Grand Canyon? Clearly some parameters need to be set.

To help answer this question let's use the specific definition of a river that cuts to any depth or length but only into strata that is still present today in the walls of Grand Canyon. (The canyon may have begun while carving into younger strata now stripped from the canyon, but ignore that for now.) For instance, the channel in Milkweed Canyon is eroded several thousand feet into the rock layers of the Grand Canyon and gravels are found at the bottom of this canyon. By our definition, this would qualify as the start of the Grand Canyon.

However, Milkweed Canyon is only a tributary to the modern Colorado River and is fully fifteen miles away from it. Is it correct to consider evidence from a faraway tributary as verification of an early incarnation of Grand Canyon, since it is difficult to know if the Milkweed drainage continued north for any length in a similar several-thousand-foot deep canyon? As the stream flowed farther away from the highlands, its gradient most likely decreased; thus its channel could have become much less deeply set within the walls of what is now the Grand Canyon. Some geologists suggest that the Esplanade could have been formed at this time as the floor of an "old" Grand Canyon (ironically, with a river going in the opposite direction of the Colorado today) but most geologists disagree on such an origin.

A few geologists have suggested that the entire length of the river within Grand Canyon owes its position to these northeast-flowing drainages. And within this relatively small group, there is at least one who thinks that the canyon was cut to its present depth at this early time! An interesting question presents itself: Has the position of the modern Colorado River system in Grand Canyon been emplaced either wholly or in part by the pattern of initial northeast drainage? The answer has yet to be determined but it is an interesting thought.

Tributary patterns in Marble Canyon may suggest that the Colorado River was positioned there during a time of northeast drainage. Tributaries are typically oriented towards the downstream flow direction of the master stream but here the opposite is seen. These tributaries enter the river towards the north, down the northeast dip of the Marble Platform but opposed to the flow of the Colorado. The trend of the river in Marble Canyon is exactly parallel to the northeast dip but mysteriously, the river today flows against this dip. If the Colorado River went north today, every aspect of this arrangement would make perfect sense. It is accepted that northeast drainage existed at an earlier time across other parts of the southern plateau and that a period of drainage reversal occurred for all or part of the Colorado River. Therefore, it may be that the Marble Canyon reach of the Colorado preserves a visible relic of ancient northeast flow in the Grand Canyon region. Geologists who favor this idea do not say that Marble Canyon was created to its present depth at this early time; they say only that the modern pattern of drainage may have been emplaced when the river flowed northeast, perhaps in strata that are today stripped back to the Echo and Vermilion cliffs. A few geologists are skeptical but it remains an intriguing hypothesis.

In any event, to what degree the Colorado River as we know it today was in existence, or to what degree the Grand Canyon was cut during this time of northeast drainage is likely to remain a mystery. The plate tectonic evidence, the scattered river deposits, and a bit of educated speculation provide an emerging picture of northeast drainage across the future Grand Canyon area, necessitating a later period of drainage reversal.

30 to 16 Million Years Ago: Initial Destruction of the Mogollon Highlands and "Confused" Drainages

The initial northeast drainage system that was established on the southern Colorado Plateau may or may not be responsible for the placement of all or part of the Colorado River in Arizona. This system may or may not have carved an early incarnation of the Grand Canyon, at least on the Hualapai Plateau. Hydrologically though, it was a relatively simple arrangement of drainage from mountains located southwest of the Grand Canyon area and rivers flowing northeast away from them. This system however, became partially disrupted beginning about 30 million years ago. Crustal disturbances in central Arizona began to affect the southern parts of the Mogollon Highlands, where the headwaters of these streams were located.

This disturbance, called the Mid-Tertiary Orogeny, was caused by stresses within the earth's crust that caused it to become stretched and thinned. The highlands floundered down somewhat from their once lofty elevations and it is likely that the streams that flowed out of them were affected by these crustal movements. As the mountains were lowered precipitation patterns probably changed towards more arid conditions. Tectonic lowering may have created catchment basins along the mountain front, disrupting the free flow of water to the northeast. There are very few deposits in the Grand Canyon area from this time period and thus it is difficult to know precisely how rivers in this area behaved. Deposits in adjacent areas however, give some clues about the changes that were occurring.

Gravels from near Sedona, Arizona document how the north-

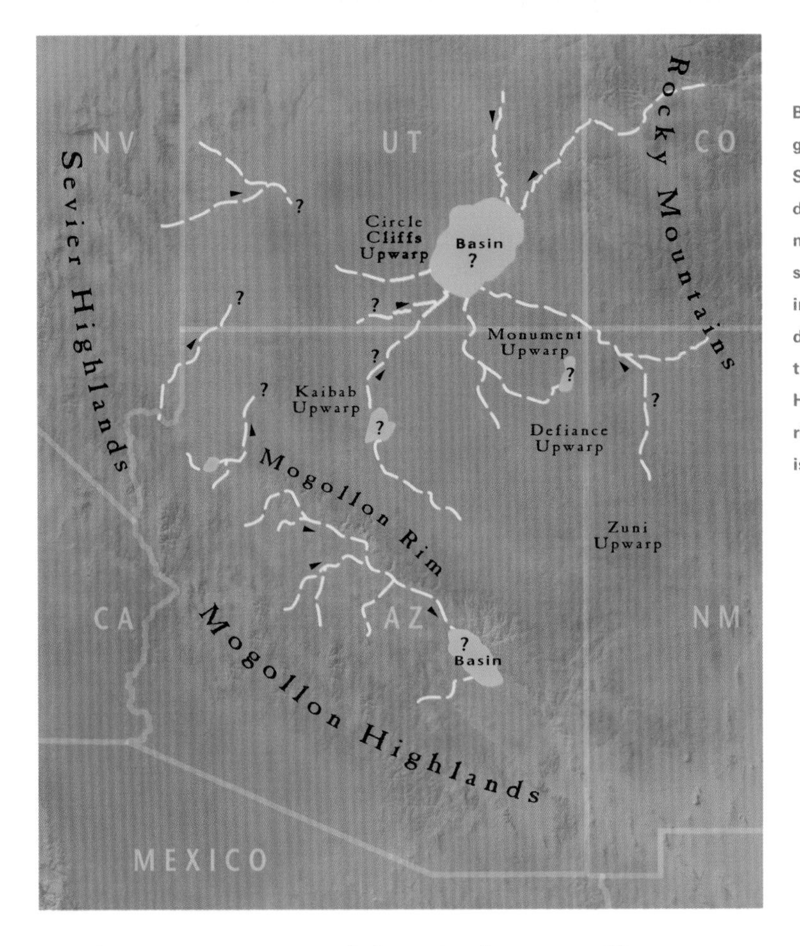

By 30 to 16 Ma, gravels exposed near Sedona, Arizona, document that northeast-flowing streams were interrupted by the development of the Mogollon Rim. How this affected rivers to the north is unknown.

east drainage was interrupted there. By about 25 million years ago streams were still flowing from the Prescott region towards the northeast, but only as far north as the Sedona area where they were deflected to the southeast at the base of the Mogollon Rim. The Rim was created by erosion as northeast-tilted strata were progressively stripped away from the uplifted flank of the Mogollon Highlands. These deflected streams flowed southeast at the base of the Rim and into some unknown destination southeast of the modern Verde Valley. Perhaps they ponded in the emerging Tonto Basin near present day Roosevelt Lake. In any case, the gravels near Sedona show that northeast drainage was disrupted here from its earlier course.

In the Grand Canyon area, only one gravel formation has been found from this entire 14-million-year time period. It is located on the Hualapai Plateau in Milkweed and Hindu canyons and is called the Buck and Doe Conglomerate. It overlies the older northeast-directed gravels but is different from them. The Buck and Doe

Conglomerate contains subangular pebbles derived mostly from rocks exposed in the canyon walls that enclose it. This suggests that the material did not travel far and probably originated in localized, not regional, streams. The Buck and Doe Conglomerate is less chemically weathered than the earlier gravels documenting that the climate was dryer than before. They are preserved within the older paleochannels carved during the period of northeast flow, but near the top this gravel deposit splays over and buries drainage divides. Its characteristics suggest to some geologists that a deeply dissected Grand Canyon was not yet in existence. The deposit is covered by a 19-million-year-old lava flow, meaning that it is at least that old and an age of 25 to 19 million years is likely.

Since there are no regional relationships recognized in Arizona among the few deposits of this age, it is difficult to establish what the overall drainage pattern may have been. The arid conditions that were developing at this time could have greatly lessened runoff towards the northeast; perhaps flow was only seasonal or even non-existent at times in the Grand Canyon region. Ponding within interior basins is a possibility as well. Or perhaps the paucity of sediments from this time represents the hazy beginning of drainage reversal for this region. It seems likely that any drainage in the Grand Canyon area at this time was "confused," meaning that it was dry, interrupted, ponded, reversed, or possibly all of the above. Only conjecture can be utilized for this enigmatic period because there is simply very little evidence for what was happening in the Grand Canyon area at this time.

16 to 6 Million Years Ago: Creation of the Basin and Range Province and Interior Drainage

As the Farallon Plate was finally consumed by the westward drift of North America, the pressure that had actively uplifted the Mogollon Highlands ended. Erosional processes became dominant in wearing down the highlands but more important, the compressional forces that had once elevated these mountains were replaced by a much different tectonic force. Extension, the stretching of the earth's crust, now began to modify those areas to the south and west of the Grand Canyon area by dropping them down into ever deepening basins, leaving the Colorado Plateau as a topographic high. Areas that were previously high mountains were downdropped into low, intermontane valleys. The Mogollon Highlands were completely destroyed during this time, which is known as the Basin and Range Disturbance. Although these events occurred outside the Grand Canyon specifically, they say something about its development since the canyon did not form in a vacuum and was created partially in response to these outside events.

Some geologists suggest that an ancient precursor to the Colorado River could have flowed through the Basin and Range and exited to the Pacific Ocean through what is now California. One theory suggests that the Los Angeles Basin may be a paleo-Colorado River delta while another geologist postulates that Monterey Bay south of San Francisco might be the mouth of such a river, only to be translated north to its present position by the San Andreas Fault. However, it seems more probable that drainage in much of the Grand Canyon area continued to be directed towards the northeast at this time, since Colorado River deposits have not been found in the Basin and Range.

When the Mogollon Highlands were faulted down creating the Basin and Range, the Colorado Plateau became elevated relative to it, even though the plateau itself may not have been uplifted one inch. This is an important distinction to note since lowering of the river's base-level could be what was perceived by earlier workers as a period of plateau uplift. Some geologists have stated, "No uplift, no canyons," but recently others have refined this to say, "No lowering of base-level, no canyons." The formation of the Basin and Range, exactly in the location where the Mogollon Highlands once stood could be the single most significant event to affect the creation of the modern drainage system. It may also satisfactorily explain when drainage reversal began for the Colorado River, or at least to the portions of it that were beginning to emerge near the plateau boundary.

Important sedimentary deposits from ancient basins are located on either side of the Grand Canyon. As low valleys formed in the Basin and Range, where the Mogollon Highlands once existed, they collected sediment derived from the mountains that surrounded them. The Grand Wash Trough west of Grand Canyon is an example of this type of interior basin. On the Colorado Plateau east of Grand Canyon lies the Bidahochi Basin. It was not created by faulting and although aspects of its history are controversial, it is quite important to the Colorado River story. These interior basins were presumably enclosed, meaning that although they could have been connected hydrologically with other basins adjacent to them they had no outlet to the sea. A discussion of the evolution of these two basins is intimately related to the origin of Grand Canyon.

The Grand Wash Trough and the adjacent Grand Wash Cliffs were formed by Basin and Range faulting between 16 and 13 million years ago. Thousands of feet of the Muddy Creek Formation accumulated in this basin. This deposit contains conglomerate, siltstone, sandstone, and limestone that lie across the path of the modern Colorado River. However, it lacks anything known to be derived from the river; the sediments originated entirely from local sources in mountains to the west. The youngest part of the deposit, the Hualapai Limestone Member, is between 11 and 6 million years old. This suggests to some that the Colorado River as we know it today, was not exiting the Grand Wash Cliffs in its present location as recently as 6 million years ago. Within the state of Colorado however, there is evidence that the river was in existence and flowing towards Utah between 20 and 10 million years ago. A great deal of controversy regarding the river's origin has resulted from the conflicting dates for it both upstream and downstream from the Grand Canyon.

The Hualapai Limestone (the upper part of the Muddy Creek Formation) west of the Grand Wash Cliffs dates to 6 million years ago and is the last deposit to pre-date the modern Colorado River. Photograph by Wayne Ranney

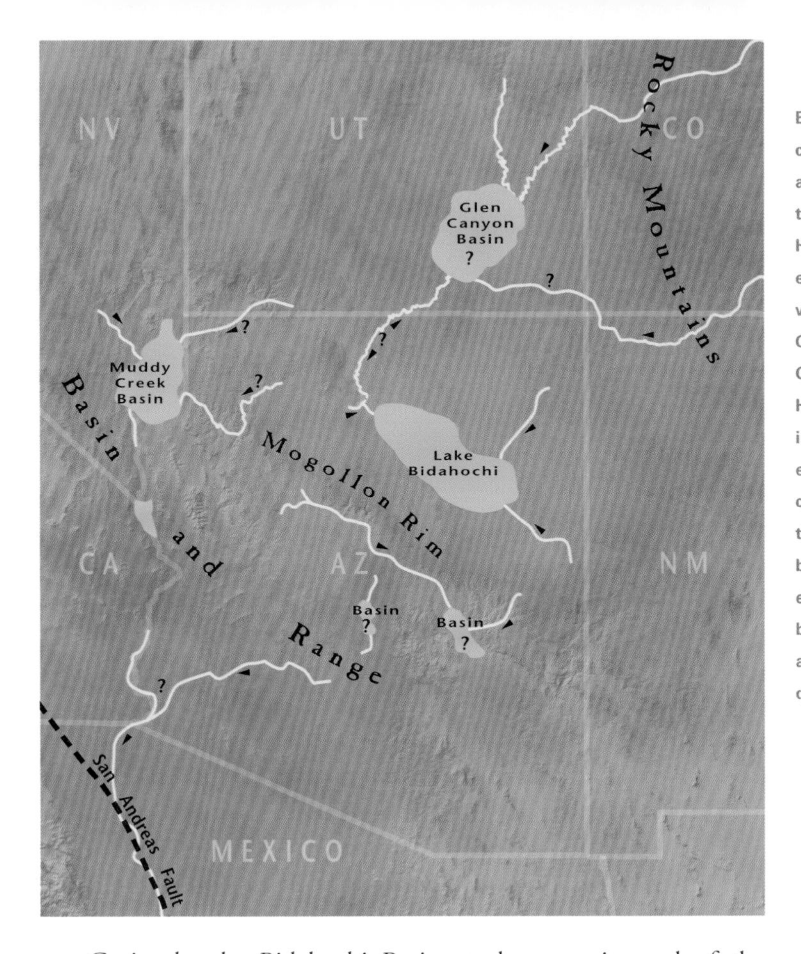

By 16 to 6 Ma, the creation of the Basin and Range destroyed the Mogollon Highlands and lowered the landscape west of Grand Canyon along the Grand Wash Cliffs. Headward erosion into the plateau edge may have commenced at this time. The Bidahochi basin was located east of Grand Canyon but the existence of a large lake here is questionable.

Curiously, the Bidahochi Basin at the opposite end of the Grand Canyon contains deposits that are exactly the same age as the Muddy Creek Formation (16 to 6 million years). However, the Bidahochi Formation is only about two hundred feet thick. A freshwater lake environment has long been invoked for these deposits and this lake has been variously called Hopi Lake (for its proximity to the Hopi Buttes Volcanic Field) or Lake Bidahochi (after the deposit). Some geologists are uneasy with the interpretation of a lake of any size or duration, since two hundred feet of sediment preserved during a 10-million-year time span seems suspect and recently the presence of such a lake has been questioned. Instead, some speculate that the Bidahochi Formation may have accumulated on an arid alluvial plain that only occasionally held small, localized ponds. As in the Grand Wash Trough, sedimentation in the Bidahochi Basin ended about 6 million years ago and these identical dates from widely separated basins hints that an event of major proportions may have occurred in the evolution of the Colorado River system. What could this event have been?

Since the 1960s, geologists have speculated that headward erosion from the Basin and Range by a young, steep-gradient river back into the Colorado Plateau may have progressed to create the Grand Canyon. In this scenario, the relative uplift of the plateau would have created the steep gradient by which this river could expand its drainage area, eventually capturing the upper Colorado River near the Kaibab upwarp. This capture event could explain why the Muddy Creek and Bidahochi formations ceased being deposited exactly 6 million years ago. Headward erosion and stream capture may have worked to integrate two separate and distinct river systems, destroying two interior basins and creating the Colorado River and the Grand Canyon in the process.

Recently, however, some geologists have voiced concerns regarding the inefficiency of headward erosion in arid climates. They note that it is highly unlikely that headward erosion could proceed almost two hundred river miles back into the Colorado Plateau rather "instantaneously," thereby connecting the Grand Wash Trough with the Bidahochi Basin at exactly the same time. Instead some of these geologists look toward a catastrophic draining of a large Lake Bidahochi that perhaps integrated the Colorado River and carved the Grand Canyon. In this scenario, as the lake filled and overtopped a low spot along its western shoreline near Grandview Point, it catastrophically drained. This could explain the small amount of Bidahochi sediment left on the present landscape, since rapid erosion and removal of the soft lake deposits might have followed the catastrophic withdrawal of lake water.

This interpretation is not without its problems as well. Much larger Ice Age lakes in the intermountain west are known to have catastrophically drained within the last thirteen thousand years and they did not carve canyons anywhere near the size or depth of Arizona's Grand Canyon. Additionally, if Lake Bidahochi did drain catastrophically, it must have followed some pre-existing line of drainage to the west, and one wonders how or when this drainage developed. Perhaps it originated by the very headward erosion that these geologists are challenging, but they do not state that. Perhaps the catastrophic draining of the lake could serve to integrate two drainages into the modern Colorado River, but taking the leap to suggest that it carved the Grand Canyon may be a stretch. At present, headward erosion and catastrophic spillover are the two major (but conflicting) ideas that geologists turn to in order to explain the origin of the modern Colorado River and the initiation of the cutting of the Grand Canyon.

Interior basin deposits on either end of the Grand Canyon are all we have to determine the course of Colorado River history between 16 and 6 million years ago. Oddly, these deposits tell us more about where the river was not flowing rather than where it was. Still, they provide clues to what landscape elements were in existence very near the canyon and could therefore tangentially suggest the history for the river and canyon. The basin deposits tell us that the Colorado River, as we know it today, did not exist as recently as 6 million years ago. What was the condition of the Grand Canyon at this time? Were segments of the modern river present on the landscape but unconnected? Were portions of the canyon present but not as deep? We are again left with a time period for which important events occurred relating to the origin of the Grand Canyon, but in which too few clues are left to say with certainty how events may have progressed.

6 Million to 2 Million Years Ago: Evidence for an Integrated Colorado River

There is solid sedimentary and tectonic evidence that rivers flowed to the northeast in the Grand Canyon area from 70 to 30 million years ago. For the next 24 million years there is very little evidence for interpreting what rivers were doing here, although the presence of interior basin deposits nearby tells us what the Colorado River was not accomplishing. Finally, after 6 million years ago, evidence appears for an integrated Colorado River that was exiting the Grand Wash Cliffs and heading generally south into the Gulf of California. The evidence for this is the presence of unequivocal Colorado River sediments in the Lake Mead area and sedimentary deposits in the lower Colorado River corridor called the Bouse Formation, and in the Salton Sea area known as the Imperial Formation. These deposits contain gravel and transported fossils that can definitely be attributed to the modern Colorado River.

The formation of the Gulf of California is one of the biggest events in the history of the Colorado River and thus the Grand Canyon. As the Farallon Plate disappeared under our continent's western edge, the relative motion between the Pacific Plate and the North America Plate gave rise to the San Andreas Fault. Through time, this fault ripped open a linear trough that was eventually inundated by water from the Pacific Ocean. The Gulf of California initially may have appeared in some form as early as 12 million years ago but evidence for its existence in the Salton Sea area suggests that it is only 6 or 7 million years old here.

When this oceanic basin was created, interior drainage basins in the Basin and Range were left differentially higher. Headward erosion from the newly formed gulf may have breached the south side of the Muddy Creek basin integrating it with drainage to the sea. Alternatively, the basin could have filled rapidly from water spilled over from Lake Bidahochi and, in turn, spilled over its own southern rim. Either of these scenarios could explain the origin of the lower Colorado River since the Bouse and Imperial formations span the change from pure Gulf of California marine deposits within them to sediments derived predominantly from the river.

The Bouse Formation is found in three separate basins astride the lower Colorado River from near Davis Dam to Yuma. A great deal of disagreement exists among geologists regarding the specific environment in which the deposit accumulated. Some exclusively marine fossils have been found within it; yet detailed chemical analysis suggests only a brackish or even a freshwater lake setting. Knowing the specific depositional environment is important because today parts of the Bouse Formation are over eighteen hundred feet above sea level and if it is marine in origin this would necessitate at least that amount of uplift in the last 4 or 5 million years. Recall from the previous section the difficulty in deciphering when uplift occurred. If the Bouse Formation was deposited in a freshwater setting, then the present elevation of the deposits could be original and no late Tertiary uplift of the plateau is required. No faults have been found bounding the Bouse deposits and the evidence

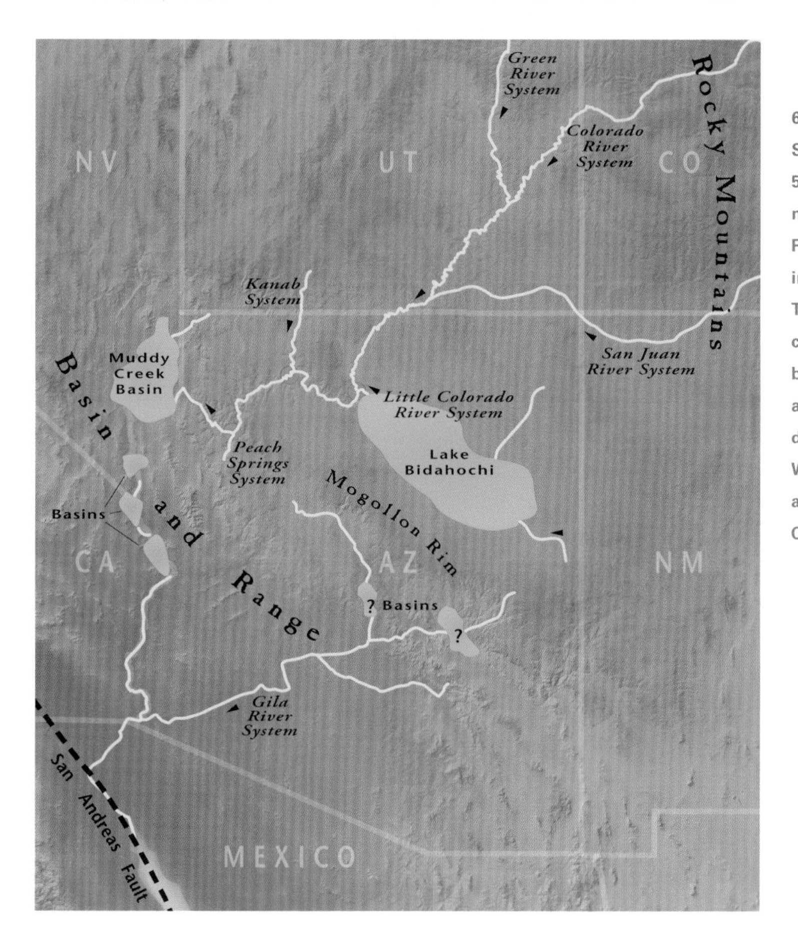

6 to 2 Ma
Sometime between 5.8 and 4.4 Ma the modern Colorado River was integrated in Grand Canyon. Three possible capture points have been proposed and are shown by the red dots at Peach Springs Wash, Kanab Creek, and the Little Colorado River.

seems to be shifting towards the freshwater interpretation. This leaves many long-held ideas about the timing of plateau uplift in doubt, causing some geologists to question the veracity of headward erosion as the driving force for integration of the Colorado River system in Grand Canyon.

The Imperial Formation is found farther south and west of the Bouse Formation and displays an upward progression from marine deposits near its base to delta deposits from the Colorado River above. It appears that the Colorado River delta grew progressively southward as material was delivered to the basin from areas farther upstream in the Grand Canyon and beyond. The formation demonstrates the increasing importance of the Colorado River in the landscape as time progressed onward from 6 million years ago.

Within the Imperial Formation are small fossils that originated in the Mancos Shale on the central Colorado Plateau. The interpretation is that these fossils were eroded and transported into the

Relative uplift or tectonic lowering?

The timing and frequency of "plateau uplift" is crucial in determining the sequence of events that carved the Grand Canyon. Uplift raised the plateau from its former position near sea level and created the high-elevation environment in which a mile-deep Grand Canyon could be cut. However, geologists have had difficulty pinpointing the specifics of the timing and frequency of plateau uplift and the controversy remains unresolved.

Geologists use clever and sophisticated techniques to determine the timing of uplift. When a highland created by uplift sheds sediments into an adjacent basin, these sediments often preserve fossils that can be dated, providing the relative age of the uplift. This method has been used to determine the timing of differential uplift in the central Rockies. In fact, the Laramide Orogeny (70 to 40 million years ago) was named from uplift-derived, fossil-rich sediments that were first described from the Laramie Basin in Wyoming.

Where basin sediments are missing however, a more complicated technique known as fission-track dating can be used to tell the age of uplift. Put simply, this method measures the presence or absence of microscopic trackways that are created by radioactive decay processes in certain minerals. Before uplift, when strata are still deeply buried, the radioactive tracks are erased by the heat present within the rocks. As the rocks are uplifted and brought closer to the surface, they begin to cool and conditions are right to preserve these fission tracks. Radiometric dates can then be obtained from within these rocks, thus dating the timing of uplift.

Imperial Formation through the Grand Canyon by an integrated Colorado River. Below a specific horizon these reworked fossils are not found; then above that horizon they are present. Reading the evidence seems to suggest that this horizon documents the moment in time when the Colorado River became fully integrated by some manner into the system we see today. This horizon is dated at approximately 5.5 million years ago. Supporting evidence comes from a lava flow near Sandy Point on Lake Mead, dated at 4.4 million years, which caps unequivocal Colorado River cobbles and gravel, giving a minimum age for an integrated river.

It is apparent that an integrated, through-going Colorado River makes its presence known on the landscape after 6 million years ago. That necessarily must be the end result of some longer landscape-forming process. Could the Grand Canyon have been in its formative stages a few million years before this date? Looking at the larger

A recent study also looked at the relative size of gas holes preserved within hardened lava flows as a way of determining the barometric pressure, and thus the elevation, of the flows at that time. Unfortunately, as sophisticated as all of these techniques have become, they continue to give conflicting answers on the timing and frequency of plateau uplift. Some geologists see evidence for as many as three discrete pulses of uplift during the Tertiary Period. Others see uplift occurring only in the early Tertiary, while others still think that late Tertiary uplift is the most important determination for when the Grand Canyon was cut. Work continues on the details of how often the plateau was uplifted but conflicting ideas could continue for many years.

However, the way that we envision uplift could possibly resolve the conflicting data. There seems to be increasing evidence against late Tertiary uplift of the Colorado Plateau. Rather, perhaps it is the *lowering* of the Basin and Range relative to the plateau edge that merely shows up in the evidence as being "plateau uplift." In this way, all plateau uplift could have occurred in the early Tertiary when the Mogollon Highlands were raised, with the tectonic lowering of the Basin and Range creating the differential elevation necessary to carve the Grand Canyon. Perhaps the late Tertiary plateau "uplift" that some geologists see evidence for is only a perceived uplift, since it is the lowering of the Basin and Range that makes it look like uplift.

picture presents some possible scenarios. Movement on the Grand Wash Fault occurred between 16 and 13 million years ago and perhaps most of the relative uplift between the Colorado Plateau and the Basin and Range was accomplished during this time. It is possible then, that headward erosion into the edge of the Colorado Plateau could have begun as early as 16 million years ago and continued during the ensuing 10 million years. It's unlikely that headward erosion would have magically begun just at the time that integration of the river occurred 6 million years ago. Perhaps the Grand Canyon was beginning to form by headward erosion during the period of interior drainage, only to be fully integrated by this process or the catastrophic spillover of lakes. Evidence for a through-going river after 6 million years ago might just be the end product of processes that began actively cutting the Grand Canyon as early as 16 million years ago.

2 Million Years Ago to the Present: Deepening of the Grand Canyon

One of the questions left unanswered in this story is when the Grand Canyon achieved its current depth. Remembering that early views tended to think of the canyon as forming quite slow and steady, we could say that it reached its current depth only yesterday! But newer studies tend to show that the canyon has become deepened in starts and stops, reacting to the influences of uplift, lowering of base-level, climate change, and runoff amounts. Perhaps as much as a third to half of the current depth of the canyon may have been accomplished in the last 2 million years—an amazing thought considering the canyon's great depth.

The last 2 million years are known as the Quaternary Period. Early geologists separated it from the Tertiary because of the evidence for a drastic climate change on planet Earth and the Quaternary Period is also known as the Ice Age. The Rocky Mountains were very much affected by the growth and decay of huge glaciers that spilled out of their valleys from the crest of the range. Certainly, lots of water was "stored" within these glaciers when they were actively forming in the Rockies.

Conditions during the Ice Age, however, were not static through time. At least four periods of glacial advance and retreat have been documented during the Quaternary. This means that as the ice melted in the Rockies, it was channeled down the Colorado River through Grand Canyon. This increased runoff must have created incredibly large floods that dwarf anything recorded in historic times. The largest flood recorded in the Grand Canyon occurred in 1957, when about 300,000 cubic feet per second raged through the gorge during the late spring melt. Geologists estimate that floods on the order of one million cubic feet per second, the average flow of the Mississippi River today, may have flowed down the Colorado during an Ice Age melt. Imagine the rocks and debris carried in such a large flood and their power to deepen the Grand Canyon!

However, other processes have also been proposed for making the Grand Canyon deeper. In western Grand Canyon, two faults cross the river and they have actively lowered the river bed by at least nineteen hundred feet on the west side of these faults. This lowering of base-

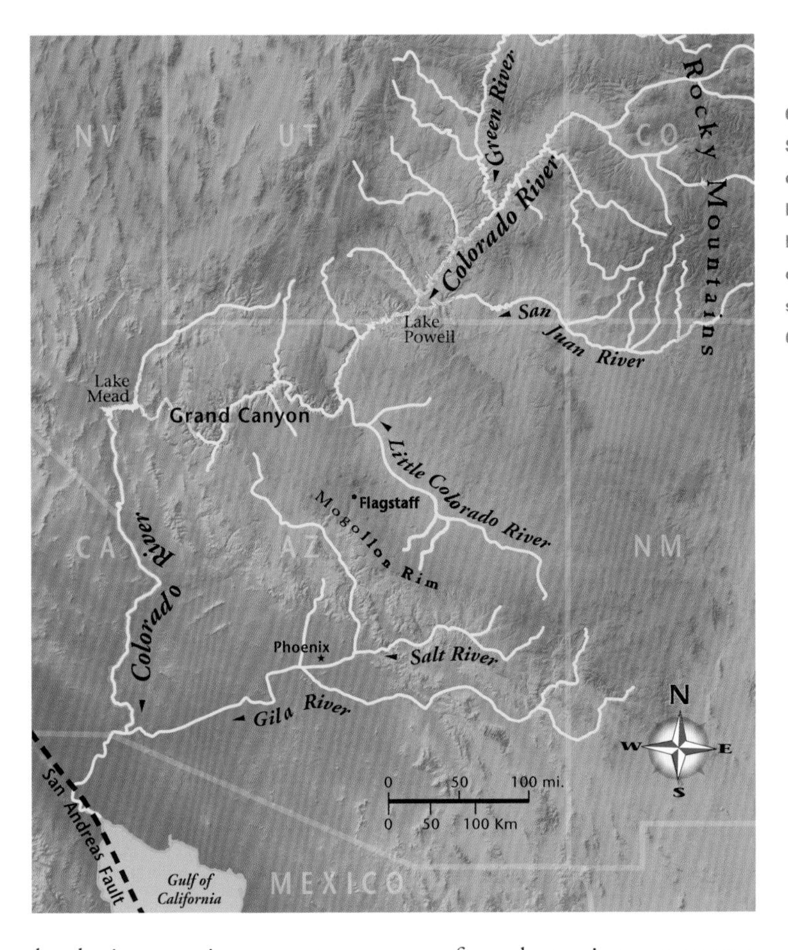

6 Ma to Present
Substantial deepening of the canyon has occurred by headward erosion or catastrophic spillover in the past 6 million years.

level is an important process for deepening upstream portions of the canyon. When an average seven-plus-magnitude earthquake occurred, it may have raised the river channel on the upthrown side of the fault, creating a knickpoint obstruction of perhaps ten feet. These knickpoints were subsequently attacked by the erosive power of the Colorado River, causing them to migrate upstream and deepen the channel of the river in the process. Together with the increased runoff in the river during the Quaternary, knickpoint migration upstream from active fault lines most likely caused the eastern Grand Canyon to become significantly deeper during the Quaternary. The Upper and Middle Granite gorges and Marble Canyon may have been deepened in this way. These three canyons are essentially large-scale slot canyons that formed quickly in response to increased runoff, lowering of base-level, and knickpoint migration.

The headwall of Hermit Canyon exposes relatively old, smooth-slopes (left and right), deeply incised by recent headward erosion (center).
Photograph by Wayne Ranney

As the river channel was deepened, the widening of the Grand Canyon must also respond in kind. Rapid deepening causes the profile of the canyon walls to become over steepened next to the river and an increase in the canyon's width probably proceeded upslope from river level towards the rim. There are a few places in Grand Canyon where it is evident that this upslope migration of canyon widening is just now reaching the rim, after many thousands or tens of thousands of years.

The last 2 million years hold even more surprising aspects concerning the canyon's spectacular evolution. Between the Toroweap and Hurricane faults in western Grand Canyon, spectacular lava flows are seen frozen in time on the canyon walls. Numerous side canyons have been filled with outpourings of lava rock called basalt. Volcanic vents producing these flows have been active within about the last 630,000 years. The lava erupted so voluminously, that huge dams were created within the Grand Canyon. One of these dams was over two thousand feet high! Another was over eighty-four miles long, although not very high in elevation. In all, at least thirteen dams may have formed within the walls of Grand Canyon. New and more reliable age dates of these flows are emerging which show that the clustering in time of the lava dams was more condensed than previously thought. Previous dates on the lava suggested they were as old as 1.8 million years, now the ages are about one third of that.

A spectacular lava cascade in a side canyon upstream from Whitmore Wash Photograph by Wayne Ranney

As spectacular as it must have been to know that this red-hot lava was pouring into a river fed by melted ice and snow from the Rockies, it is the geologic lessons learned from this lava history that is truly astounding. A two thousand-foot-high dam, if it survived quick destruction or collapse, would have created a natural reservoir on the Colorado River that would have backed up to an area near

Moab, Utah over 300 miles upstream! Below Grand Canyon Village, Indian Garden campground would be under two hundred feet of water!

Unfortunately, no reservoir deposits have been recognized in Grand Canyon and some geologists are urging caution when invoking the presence of long-lived lava dam reservoirs in the canyon. They argue that any lava dam may have been inherently unstable because of the presence of unconsolidated river gravel at the base of the dam, or the fact that the lava may have become shattered or glassy as it entered the cold water. This possibility suggests that some lava dams may have been short-lived phenomena. The rapid failure of a lava dam could explain why no reservoir deposits have been preserved. Conversely, the lack of these deposits could be explained by late Quaternary deepening of the canyon which would remove any reservoir deposits as the canyon became deeper and wider.

Remarkably, even if some lava dams existed for hundreds or thousands of years, they had to be removed before the next one took its place, since dam buttresses are inset one against the next. This speaks clearly to the enormous cutting power of the Colorado River. Lava dams upset the equilibrium of the river's gradient and rivers despise these perturbations in gradient. If a lava dam existed long enough to fill with water, imagine the scene as water poured over the top or through cracks and weaknesses in the dam. Erosional removal of these dams, if water poured over the top, most likely occurred quite rapidly geologically speaking. Another outpouring of lava started the process all over again.

After the 2000 symposium it was reported that at least five of the lava dams were destroyed catastrophically, perhaps even instantaneously, due to instability of the dam. Huge outburst flood deposits have been recognized along the river corridor below the lava dams. These deposits consist mostly of angular basalt rock; one of the deposits was found six hundred fifty feet above the river and a boulder in it was over one hundred feet in diameter. Imagine the power and size of the floods that are able to move material of this size high up on the canyon walls. Age dates of these outburst flood deposits show that catastrophic failures occurred between 525,000 and 100,000 years ago.

Lava cascades created dams behind which reservoirs may have formed. Some dams may have failed before filling their reservoirs with water; others probably overflowed, eroding the dam's unstable foundation.

Picture for a moment times past in the Grand Canyon. On a cold winter's night, red rivers of lava poured over the canyon rim, cascading into the icy Colorado and filling the canyon with hot steam. Imagine seeing the downstream channel run dry as water becomes trapped behind a lava dam. Eventually, if the dam is strong and durable, waterfalls pour over them, creating cascades as high as eight hundred, twelve hundred, or even two thousand feet, all set within the colorful walls of the Grand Canyon. Imagine the view from the rim at the moment when a lava dam catastrophically failed and a tremendous outburst flood roared through the Grand Canyon with rubble-filled water over six hundred feet deep. The sight and sounds must have been phenomenal and these new discoveries give us a window to witness the phenomenal events that once occurred here. The Grand Canyon has certainly been an exciting place in the last 630,000 years!

Remnants of lava dams still cling to canyon walls. Photograph by Bronze Black

Exciting as these scenes are, it is the lessons these lava dams teach us that allow us to learn something about the canyon-forming process. It's a curious fact that when a lava dam was created, then removed, erosion proceeded precisely to the pre-dam profile of the canyon before the next flow was emplaced. This tells us that the depth and profile of the canyon walls must be in some sort of equilibrium with the river. When this equilibrium is upset by something like a lava dam or faulted knickpoint, the river responds instantaneously in an attempt to recover its former "comfortable" profile. Once this profile has been attained, the river does not continue to deepen or widen the canyon any further. Some geologists use this evidence to say that the canyon may experience discrete periods of active deepening and widening (in response to volcanism, uplift, or climate change), separated by periods when the canyon just sits there and doesn't change very much.

An interesting story is beginning to emerge. The canyon may have been significantly deepened within the last 2 million years. Yet, it may undergo significant and longer periods of time when it just sits there, not getting appreciably deeper or wider. When someone asks, Is the Grand Canyon getting deeper? perhaps we should appropriately respond, Yes and no, or Maybe not now but certainly in the future.

SUMMARY

The Grand Canyon area was at sea level 80 million years ago but was progressively uplifted. Earlier rivers flowed to the northeast, opposite the flow direction of the Colorado River today. At 30 million years ago the landscape had relatively low relief and was at an unknown elevation. Creation of the Basin and Range 16 million years ago and the Gulf of California by about 6 million years ago allowed separate river systems to become integrated, either by headward erosion and stream piracy, or the catastrophic spillover of ancient lakes. Today, the Grand Canyon is a deeply incised landscape with a high average elevation.

6

Summary

BY NOW, we may begin to appreciate the difficulty in knowing the precise history and evolution of the Grand Canyon. It may appear to be an impossible task to look back through so much time and so much erosion and definitively piece together a coherent story of how this great gorge was carved. And when we consider all of the variations that are certainly plausible, we see that there are many possibilities that branch backwards in myriad directions like the genealogy of a family tree. We find that if we follow any one branch of that tree far enough back in time, it takes us farther and farther away from other branches that are also possible and no less related to the ultimate descendent. For these reasons, a definitive story of how the Grand Canyon was carved may forever remain unknown to us.

However, in almost one hundred fifty years of scientific study we have learned a few "truths" about the canyon's formation and it may be possible to string together a few of these accepted facts into a satisfactory explanation. A word exercise can be used to help us understand this complex history in an easy way. In this exercise we are given a chance to tell the story of Grand Canyon's origin but the challenge is to tell the greatest "truth" using the fewest words. As we are given permission to use more words, our explanation can expand but we must still say something that is agreed

The Colorado River flows beneath Toroweap Overlook into Lava Falls in western Grand Canyon. Photograph by Bob and Sue Clemenz

upon by most geologists without veering off too much into the "what-ifs" or "maybes."

To begin, we are given just one sentence to say how the Grand Canyon may have formed. What one sentence tells the most truth about Grand Canyon's origin? This sentence might read:

The Grand Canyon was carved by the Colorado River.

This is the simplest and most truthful way to explain Grand Canyon's origin because all geologists agree on the close relationship between the history of the river and that of the canyon. We could stop right here with this one sentence and some people would be completely satisfied to know the Grand Canyon was formed by the down-cutting of the Colorado River. But for those who want to know a bit more, what one additional sentence could we add to our simplified explanation of Grand Canyon's history that would tell the greatest truth? It might read:

To date, geologists have been unable to determine the canyon's precise age and what specific processes were at work in carving it.

Now we introduce the important idea that there are unknown's in our attempt to decipher this story. We have explained that a precise age of the canyon still eludes us and that there are many possibilities that could have been at work in making the Grand Canyon. This simple truth frees us to explore the broader concepts in the canyon's development rather than be tied into the specific details, which in many instances have only served to confuse us. Is there another sentence that can be added at this point that would speak the greatest truth? Let's try:

In fact, the many geologists who have devoted their careers to studying the canyon cannot resolve its age more precisely than somewhere between 80 and 6 million years old, and they still debate whether it was formed rather catastrophically or over a much more extended period of time.

Now we know just how broad the parameters are in our understanding of the canyon. If we accept that every idea ever proposed contains at least some grain of truth within it, then this last sentence serves us well to know what the extreme limits of the canyon's age may be and the variety of processes that are possible in its formation. At this point in the exercise we are allowed to add two sentences to our growing list of truths. This allows us to know something about the larger history of the region that is accepted by the greatest number of geologists:

All geologists agree that the sea last withdrew from the Grand Canyon region about 80 million years ago, leaving the low-lying, subdued landscape upon which an initial river system was established. Ironically, during the ensuing 50 million years, these rivers flowed northeast in the Grand Canyon region, directly opposite the direction of the modern Colorado River.

The larger geologic study of the western United States reveals evidence for this truth. Central Arizona and Nevada were areas of highlands where rivers drained north and east towards lowlands that today are part of the Colorado Plateau. Few, if any, geologists would debate this "fact" since gravel has been found in the Grand Canyon region that shows a northeast flow direction and lake deposits of the same age are widespread to the north. The exact same geologic setting is present today on the eastern slope of the Andes Mountains where the headwaters of the Amazon are located. The next sentence adds to these thoughts, and is an important qualifier to the puzzle.

It is still unknown whether these northeast-flowing streams are responsible for the present position and configuration of all or part of the Colorado River in Grand Canyon. However, during the period from 70 to 40 million years ago, the area that is now the Colorado Plateau was uplifted, perhaps to its present-day elevation, and it could be that some portions of the river's course in Grand Canyon were firmly entrenched at this time.

A moderately deep tributary in the modern Grand Canyon contains gravels deposited by a northeast flowing stream, but nowhere are these found in the canyon proper. In other words, we can't say for certain if the Colorado River was positioned by these northeast-flowing rivers or if the Grand Canyon was at least partially cut at this time. What we do know is that the plateau was elevated considerably at this time, setting the stage for future deepening of the canyon. What follows?

About 30 million years ago, the source area for the northeast-flowing streams began to subside through faulting and erosion. The effect this had on these rivers is unknown because no deposits have been found. These rivers may have become "confused," ephemeral, ponded, reversed, or relocated during this time. We simply don't know. Rivers in the area "disappear" from our view from between 30 and 16 million years ago.

We are, of course, on increasingly shaky ground. The time period between 30 and 16 million years ago reveals the least evidence for what might have occurred in the Grand Canyon area. Virtually no sediment is left to tell us what happened during this time. This in itself may serve as crude evidence. Perhaps the Colorado Plateau was eroding at this time and not accumulating sediment. Or alternatively, perhaps the rivers on the plateau were too diminished or compromised to leave much sediment in their wake. Perhaps both. We will probably never know very much from this time period. Did the rivers still drain the plateau in a northeast direction into a basin of interior drainage? Or did they find a route to the southwest across the destroyed highlands and into the Pacific Ocean? We might have a possible answer when we realize that:

Interior drainage basins existed on the Colorado Plateau before 30 million years ago and in the adjacent Basin and Range Province after about 16 million years ago, which may hint at what was present in the Grand Canyon region during this enigmatic time of drainage reversal.

Interior drainages, those with no outlet to the sea, seem to be prevalent for much of the history of the Colorado River. Evidence for interior drainage, both before and after this time period, might suggest that interior drainage existed across the whole spectrum of time and during the period of drainage reversal. We know fairly well what happened next:

Beginning 16 million years ago, the old Mogollon Highlands were completely destroyed by faulting, which differentially separated the Colorado Plateau from the Basin and Range. An interior basin deposit called the Muddy Creek Formation accumulated west of the plateau edge but does not contain obvious evidence for the presence of the Colorado River. East of the Grand Canyon, the Bidahochi Formation may record the site of a lake environment. Both of these basins were breached sometime after 6 million years ago when the San Andreas Fault created the Gulf of California and drainage was directed off the southwestern edge of the plateau. The integration of multiple river systems by headward erosion, catastrophic spillover of lakes, or a combination of the two, is what most likely created the Colorado River we see today.

The subsidence of the Basin and Range between 16 and 6 million years ago caused a tremendous difference in elevation between it and the Colorado Plateau. This relief may have caused water to be directed off the plateau towards the southwest but there is very little evidence for this. Interior drainage was probably occurring on the plateau and within the Basin and Range at this time. With the opening of the Gulf of California about 6 million years ago, the interior drainage of the Basin and Range and the Colorado Plateau was breached towards the sea and the modern Colorado River was born. Finally:

Recent studies suggest that the Grand Canyon may have become significantly deeper in the last 3.5 million years. This was accomplished by the elevation difference between the plateau edge and the Basin and Range (causing the gradient of the Colorado River to become steepened), the upstream deepening of the canyon after movement on faults, and increased runoff during the Ice Age. Spectacular lava flows cascaded into the Grand Canyon within the last 630,000 years and created lava dams and perhaps reservoirs behind them. Huge outburst floods have recently been documented showing that some of these lava dams failed catastrophically. These events reveal the catastrophic nature of Ice Age deepening of the Grand Canyon.

Now we have a relatively short narrative that summarizes the possible origin and development of the Grand Canyon and the Colorado River. It serves as a concise story about the history of the river and the canyon.

Concise Summary

The Grand Canyon was carved by the Colorado River. To date, geologists have been unable to determine the canyon's precise age and what specific processes were at work in carving it. In fact, the many geologists who have devoted their careers to studying the canyon cannot resolve its age more precisely than somewhere between 80 and 6 million years old, and they still debate whether it was formed rather catastrophically or over a much longer period of time.

All geologists agree that the sea last withdrew from the Grand Canyon region about 80 million years ago, leaving the low-lying, subdued landscape upon which an initial river system was established. Ironically, during the ensuing 50 million years, these rivers flowed northeast in the Grand Canyon region, directly opposite the direction of the modern Colorado River. It is still unknown whether these northeast flowing streams are responsible for the present position and configuration of all or part of the Colorado River in Grand Canyon. However, during the period from 70 to 40 million years ago, the area that is now the Colorado Plateau was uplifted, perhaps to its present-day elevation, and it could be that some portions of the river's course in Grand Canyon were firmly entrenched at this time.

About 30 million years ago, the source area for the northeast-flowing streams began to subside through faulting and erosion. The effect this had on these rivers is unknown because no deposits have been found. These rivers may have become "confused," ephemeral, ponded, reversed, or relocated during this time. We simply don't know. Rivers in the area "disappear" from our view from between 30 and 16 million years ago. Interior drainage basins existed on the Colorado Plateau before 30 million years ago and in the adjacent Basin and Range Province after about 16 million years ago, which may imply something about what was present in the Grand Canyon region during this enigmatic time of drainage reversal.

Beginning 16 million years ago, the old Mogollon Highlands were completely destroyed by faulting, which differentially separated the Colorado Plateau and the Basin and Range. An interior basin deposit called the Muddy Creek Formation accumulated west

Detail from *Point Sublime* by William Henry Holmes, 1882

of the plateau edge but does not contain evidence for the presence of the Colorado River. East of the Grand Canyon, the Bidahochi basin may have been the site of a lake environment. Both of these basins were breached sometime after 6 million years ago when the San Andreas Fault created the Gulf of California and drainage was directed off the southwestern edge of the Plateau. The integration of multiple river systems by headward erosion, catastrophic spillover of lakes or a combination of both, is what most likely created the Colorado River we see today.

Recent studies suggest that the Grand Canyon may have become significantly deeper in the last 3.5 million years. This was accomplished by the elevation difference between the Plateau edge and the Basin and Range (causing the gradient of the Colorado River to become steepened), the upstream deepening of the canyon after movement on faults, and increased runoff during the Ice Age. Spectacular lava flows cascaded into the Grand Canyon within the last 630,000 years and created lava dams and perhaps reservoirs behind them. Huge outburst floods have recently been documented showing that some of these lava dams failed catastrophically. These events reveal the catastrophic nature of Ice Age deepening of the Grand Canyon.

GLOSSARY

antecedence - a stream that flowed in its present course prior to the development of the existing topography
anticline - a fold that is convex upward such that it forms an arch in rock strata
barbed tributaries - tributary streams that flow against the direction of some larger drainage, suggesting drainage reversal of the larger stream
basalt - hardened lava with about 50 percent silica
basin - the depressed drainage or catchment area of a stream or lake
Basin and Range - a physiographic region west of the Colorado Plateau characterized by a series of parallel mountains and valleys and created about 16 million years ago
bedrock - referring to any consolidated rock that underlies the rocks or sediment being described
carbonate - compound of carbon and oxygen, typically found as limestone or dolomite
Cenozoic - the most recent geologic era from 65 million years ago to the present
clast - an individual fragment or grain in a sedimentary rock formed by the disintegration of some larger rock
Colorado Plateau - the region of the Four Corners states that was gradually uplifted such that the sedimentary rocks remain nearly horizontal
compression - a system of forces or stresses that tend to shorten but thicken rocks
conglomerate - rounded, water-worn fragments of pebbles cemented together by other minerals
consequent stream - one which follows a course that is a direct consequence of the original slope on which it developed
contact - the place or surface where two different kinds of rocks come together
cross-bed - laminations of strata that are oblique to the main planes of stratification
crust - the outer rigid layer of the earth which floats on the earth's mantle
depositional environment - the surface conditions such as geographic setting, climate, and transport medium that affects the nature of sedimentary deposits
differential uplift - where one part of the earth's crust rises more or faster than that adjacent to it
dip - the angle at which a stratum or any planar feature is inclined from the horizontal
drainage - any area where water is removed by down-slope flow
drainage divide - the topographic border between adjacent drainage basins or watersheds
drainage reversal - the process whereby a river changes its flow direction
ephemeral - an intermittent flowing stream
equilibrium - when the phases of any system do not undergo any change in property with the passage of time
era - a division of geologic time of the highest order
erosion - the group of processes whereby rock is loosened or dissolved and moved from any part of the earth's surface
escarpment - a steep face terminating a high surface; a cliff
extension - the pulling apart of the earth's crust
facies - the general appearance of a rock body as contrasted with other parts
fault - a fracture along which there has been displacement of the earth's crust
floodplain - the portion of a river valley

that is flooded during excessive runoff
fold - a bend in strata
formation - the primary unit of rock layers
granite - a plutonic rock rich in silica
headward erosion - where a stream lengthens its valley and is cut back by the water which flows in at its head
Holocene - a geologic epoch consisting of the last 10,000 years
interior basin - a valley, usually with sedimentation, having no outlet to the sea
knickpoint - points of abrupt change in the longitudinal profile of a stream
lacustrine - pertaining to lakes and lake environments
Laramide Orogeny - mountain building episode from 80 to 40 million years ago that raised the Colorado Plateau
lava - molten rock of igneous origin erupted onto the earth's surface
limestone - sedimentary rock composed mostly of the mineral calcite
lithology - pertaining to rock type, i.e. sandstone, shale, etc.
mantle - layer of the earth's interior below the crust and above the core
marine - pertaining to the sea or oceanic environment
mature - landscapes in which maximum development has been reached
Mesozoic - the third era of geologic time lasting from about 245 to 65 million years ago
Mid-Tertiary Orogeny - a crustal disturbance located in central Arizona and lasting from about 30 to 20 million years ago
Mogollon Highlands - a mountain range present southwest of Grand Canyon from about 80 to 16 million years ago
monocline - a one-limbed flexure in structure in which the beds are flat lying except the flexure itself
mudstone - a fine-grained sedimentary rock that includes clay, silt and sand clasts
obsequent - a stream flowing in the opposite direction of the dip of the strata or the tilt of the surface
orogeny - a mountain-building event
outcrop - the exposure of bedrock or strata projecting out from the overlying cover of soil or detritus
paleogeography - the study of geography through geologic time
Paleozoic - the second era of geologic time lasting from about 540 to 245 million years ago
plate tectonics - a theory of global-scale dynamics involving the movement of many rigid plates of the earth's crust
Pleistocene - an epoch of the Quaternary time period lasting from about 2 million years ago to 10,000 years ago
Quaternary - the most recent geologic time period lasting from about 2 million years ago to the present
relief - the range of topographic elevation within a specific area
rift - a deep fracture or break in the earth's crust where two plates pull apart
Rocky Mountains - a physiographic province that was uplifted during the Laramide Orogeny
sandstone - a sedimentary rock containing a large quantity of quartz sand grains
sapping - where groundwater flows out of the ground and caused rocks to become undercut
scarp retreat - gradual retreat of a cliff face by erosion
shale - a fine-grained sedimentary rock composed of clay and silt grains that splits readily into thin layers

sorting - a descriptive term used to indicate the degree of similarity in a sediment with respect to the size of the grains

spreading center - the place in the ocean where rifts form in the earth's crust

strata - the layering found within sedimentary rocks regardless of thickness

stream piracy - process by which the headwaters of a steeper stream erode headward and capture a lower gradient stream

subangular - a measure of roundness in gravel in which definite effects of wear are shown; fragments retain their original form and the faces are virtually untouched, but the edges and corners are rounded off to some extent.

subduction - tectonic process in which a dense oceanic plate dives beneath another piece of crust due to plate convergence

subsequent streams - those that have grown longitudinally along belts of soft strata

superposed streams - those that were emplaced on a new surface and that maintained its course as it eroded down into pre-existing structures

syncline - a fold that is concave upward such that it forms a sag in rock strata

Tertiary - the earliest time period of the Cenozoic Era lasting from about 65 to 2 million years ago

unconformity - a surface where there is a gap in the rock record

uplift - the elevation of a part of the earth's crust

upwarp - a specific area that has been uplifted

SCIENTIFIC BIBLIOGRAPHY

Babenroth, Donald L. and Strahler, Arthur N., 1945. Geomorphology and structure of the East Kaibab monocline, Arizona and Utah: GSA Bulletin, v. 75, #9, p. 107-150.

Blackwelder, Eliot, 1934. Origin of the Colorado River: GSA Bulletin, v. 45, # 3, p. 551-566.

Cooley, Maurice E. and Davidson, E. S., 1963. The Mogollon Highlands — Their influence on Mesozoic and Cenozoic erosion and sedimentation: Arizona Geological Society Digest, v. 6, p. 7-35.

Davis, William Morris, 1901, An excursion to the Grand Canyon of the Colorado; Harvard College, Bulletin of the Museum of Comp. Zoo., v. 38, Geological series 5, p. 107-201.

Dumitru, Trevor A. and others, 1994. Mesozoic-Cenozoic burial, uplift, and erosion history of the west-central Colorado Plateau: Geology, v. 22, p.499-502.

Dutton, Clarence E., 1882. Tertiary history of the Grand Ca–on district: USGS Monograph 2, 264 p.

Elston, Donald P. and Young, Richard A., 1991. Cretaceous-Eocene (Laramide) landscape development and Oligocene-Pliocene drainage reorganization of Transition Zone and Colorado Plateau, Arizona: Jour. of Geophys. Res., v. 96, #87, p. 12,389-12,406.

Emmons, S. F., 1897, The origin of the Green River; Science, v. 6, # 131, p. 20-21.

Fenton, Cassandra R. and others. 2001, Displacement rates on the Toroweap and Hurricane faults: Implications for Quaternary downcutting in the Grand Canyon, Arizona: Geology, v. 29, #11, p. 1035-1038.

Goldstrand, Patrick M., 1994. Tectonic development of upper Cretaceous to Eocene strata of southwestern Utah: GSA Bulletin, v. 106, p. 145-154.

Gregory, Herbert E., 1917. Geology of the Navajo Country: A reconnaissance of parts of Arizona, New Mexico, and Utah: USGS Prof. Paper 93.

Gregory, Herbert E., 1947. Colorado River drainage basin: Am. Jour. of Sci., v. 245, # 11, p. 694-705.

Hamblin,W. K., 1994. Late Cenozoic lava dams in the western Grand Canyon: GSA Memoir 183. 139 p.

Holm, Richard F., 2001. Cenozoic paleogeography of the central Mogollon Rim-southern Colorado Plateau region, Arizona, revealed by Tertiary gravel deposits, Oligocene to Pleistocene lava flows, and incised streams: GSA Bulletin, v. 113, p. 1467-1485.

Hunt, Charles B., 1956. Cenozoic geology of the Colorado Plateau: USGS Prof. Paper 279, 99 p.

Hunt, Charles B., 1969. Geologic history of the Colorado River: USGS Prof. Paper 669, p. 59-130.

Ives, Lieutenant Joseph Christmas, 1861. Report upon the Colorado River of the West, US 36th Cong., 1st Session, Senate Exec. Doc., Washington D. C., p. 93-112.

Johnson, Douglas Wilson, 1909. A geological excursion in the Grand Canyon district: Boston Soc. Nat. Hist. Proceedings, v. 34, p. 135-161

Koons, Donaldson, 1948. Geology of the eastern Hualapai Reservation: Plateau, v. 20, #4, p. 53-60.

Larsen, Edwin E., 1975. Late Cenozoic basic volcanism in northwestern Colorado and its implications concerning tectonism and the origin of the Colorado River system: GSA Memoir 144, p. 155-178.

Lee, Willis T., 1906. Geologic reconnaissance of a part of western Arizona: USGS Bulletin 352, 96 p.

Longwell, Chester R., 1946. How old is the Colorado River: Am. Jour, Sci., v. 244, #12, p. 817-835.

Lovejoy, Earl M. P., 1980. The Muddy Creek Formation at the Colorado River in Grand Wash: The dilemma of the immovable object: Az. Geol. Soc. Digest, v. 12, p.177-192.

Lucchitta, Ivo, 1989. History of the Grand Canyon and of the Colorado River in Arizona: Az Geol. Soc. Digest, v. 17, p. 701-716.

Lucchitta, Ivo, 1979. Late Cenozoic uplift of the southwestern Colorado Plateau and adjacent lower Colorado River region: Tectonophysics, v. 61, p. 63-95.

McDougall, Kristin and others, 1999. Age and paleoenvironment of the Imperial Formation near San Gorgonio Pass, So. California: Jour. of Foram. Res., v. 29, #1, p 4-25.

McKee, Edwin D. and others, 1967. Evolution of the Colorado River in Arizona: MNA Bulletin # 44, 67 p.

McKee, Edwin D., 1972. Pliocene uplift of the Grand Canyon region - Time of drainage adjustment: GSA Bulletin, v. 83, p. 1923-1931.

Newberry, John Strong, 1861. Report upon the Colorado River of the West, US 36th Cong., 1st Session, House Exec. Doc. 90, pt. 3, 154 p.

Pederson, Joel, 2002. Colorado Plateau uplift and erosion evaluated using GIS: GSA Today, v. 12, # 8, p. 4-10.

Pederson, J.L., Schmidt, J.C., and Anders, M.D., 2003, Pleistocene and Holocene geomorphology of Marble and Grand Canyons, canyon cutting to adaptive management: in Eastbrook D.J. ed., Quaternary Geology of the United States, INQUA 2003 Field Guide Volume, Desert Research Institute, Reno, NV. P. 407-438.

Peirce, H. Wesley, 1979. An Oligocene (?) Colorado Plateau edge in Arizona: Tectonophysics, v. 61, p. 1-24.

Potochnik, Andre R. and Faulds James E., 1998. A tale of two rivers: Tertiary structural inversion and drainage reversal across the southern boundary of the Colorado Plateau: GSA Field Trip Guidebook, Rocky Mt. Sec. Meeting.

Powell, John Wesley, 1875; Exploration of the Colorado River of the West and its Tributaries, Smithsonian Institution Annual Report, 291 p.

Ranney, Wayne D. R., 1988. Geologic history of the House Mountain area, Yavapai County, Arizona: M.S. thesis, Northern Arizona University, Flagstaff, 99 p.

Ranney, Wayne D. R., 1991. Mid-to late-Tertiary evolution of the Mogollon Rim near Sedona, Arizona: GSA Abstracts with Programs, Rocky Mountain. and South-Central Sections, v. 23, # 4, p. 58.

Ranney, Wayne D. R., 1998. Geomorphic evidence for the evolution of the Colorado River in the Little Colorado-Marble Canyon area, Grand Canyon, Arizona: GSA Abstracts with Programs, Rocky Mountain Section, v. 30, # 6, p. 34.

Robinson, H. H., 1910. A new erosion cycle in the Grand Canyon District, Arizona: Jour. Geology, v. 18, # 8, p. 742-763.

Scarborough, Robert B., 1989. Cenozoic erosion and sedimentation in Arizona: Az Geol. Soc. Digest, v. 17, p. 515-538.

Schmidt, Karl-Heinz, 1989. The significance of scarp retreat for Cenozoic landform evolution on the Colorado Plateau: Earth Surface Proc. and Landforms, v. 14, p. 93-105.

Spencer, Jon E. and Patchett, P. Jonathan, 1997. Sr isotope evidence for a lacustrine origin for the upper Miocene to Pliocene Bouse Formation, lower Colorado River trough, and implications for timing of Colorado Plateau uplift: GSA Bulletin, v. 109, p. 767-778.

Stock, Chester, 1921. Late Cenozoic mammalian remains from the Meadow Valley region, southeastern Nevada: Amer. Jour. of Sci., v. 5, #2, p. 250-264.

Strahler, Arthur N., 1948. Geomorphology and structure of the West Kaibab fault zone and Kaibab Plateau, Arizona: GSA Bulletin, v. 59, #6, p. 513-540.

Walcott, C. D., 1890; Study of a line of displacement in the Grand Cañon of the Colorado, in northern Arizona; GSA Bulletin, v. 1, p. 49-64.

Young, Richard A. and Brennan, William J., 1974. Peach Springs tuff: Its bearing on the structural evolution of the Colorado Plateau and development of Cenozoic drainage in Mohave County, Arizona: GSA Bulletin, v. 85, p. 83-90.

Young, Richard A. and McKee, Edwin H., 1978. Early and middle Cenozoic drainage and erosion in west-central Arizona: GSA Bulletin, v. 89, p. 1745-1750.

Young, Richard A. and Spamer, Earl E., eds., 2004. The Colorado River: Origin and Evolution; Grand Canyon Association Monograph no. 12, 280 p.

EARLIER POPULAR BOOKS CONCERNING GRAND CANYON'S ORIGIN

Darton, Nelson Horatio, 1917. Story of the Grand Canyon of Arizona: Fred Harvey Company. 75 p.

Maxson, John H., 1962. Grand Canyon: Origin and scenery: Grand Canyon Natural History Assn. 35 p.

McKee, Edwin D., 1931. Ancient landscapes of the Grand Canyon region: Northland Press. 52 p.

ABOUT THE AUTHOR

Wayne Ranney is a geologist, educator, and guide who became interested in Earth history while working as a backcountry ranger at Phantom Ranch at the bottom of the Grand Canyon from 1975 to 1978. Since that time, his life has revolved around the canyons, rivers, and red rock stratigraphy found in the American Southwest. He received both his bachelor's and master's degrees from Northern Arizona University in his hometown of Flagstaff, Arizona. He enjoys international and domestic travel, hiking, rafting, photography, watching the sky, and has plans to design his own home.

Wayne is an adjunct professor of geology at Yavapai College in Sedona, Arizona, and leads field trips throughout the southwest for such diverse organizations as the Grand Canyon Field Institute, the Museum of Northern Arizona, the American Orient Express, Smithsonian Journeys, and TCS Expeditions. He is a popular lecturer on a variety of international expeditions and has visited and lectured on all seven of the earth's continents. Wayne is a contributing science writer for *Sedona Magazine* and is the author of *Sedona Through Time: Geology of the Red Rocks*. He enjoys leading educational excursions to many of our planet's most interesting landscapes. You can visit his Web site at: **www.wayneranney.com**

INDEX